Oliver Perry Hay

The batrachians and reptiles of the state of Indiana

Oliver Perry Hay

The batrachians and reptiles of the state of Indiana

ISBN/EAN: 9783337306021

Printed in Europe, USA, Canada, Australia, Japan

Cover: Foto ©Andreas Hilbeck / pixelio.de

More available books at **www.hansebooks.com**

THE

BATRACHIANS AND REPTILES

—— OF THE ——

STATE OF INDIANA.

BY

OLIVER PERRY HAY, A. M., Ph. D.

INDIANAPOLIS:
WM. B. BURFORD, PRINTER AND BINDER.
1892

PREFACE.

The Batrachia and Reptilia of the State of Indiana form the subjects of the following pages. In the body of this work I have endeavored to include all the species known to inhabit the State, and to exclude all that are not known to occur within our limits. Since, however, there are several species which are, judging from their geographical distribution, likely hereafter to be taken within the State, I have added their names in an appendix.

Of all the species mentioned in this paper I have given as accurate descriptions as I have been able to prepare; and I have endeavored to state also the most important facts known regarding their habits. It is to be hoped that this endeavor will incite others to study our lower vertebrates with respect to their manner of life, since too little is known about even the commonest species.

I am indebted to many friends for aid in preparing this work, so many that I can not here mention all their names. Under each species I have tried to give due credit for specimens and notes. I must here, however, acknowledge the liberality of Dr. Leonhard Stejneger, Curator of Reptiles in the National Museum, Washington, D. C., in giving me free access to the large collections there; also the kindness of Prof. B. W. Evermann, who allowed me to examine a considerable collection made by himself, mostly in the vicinity of Terre Haute. This collection is the property of the State Normal School. To Prof. S. S. Gorby I owe the opportunity to examine the specimens in the State Museum. Some years ago I had occasion to study a collection made at New Harmony by the late Mr. James Sampson, of that place. For the opportunity of doing this I am indebted to Prof. John Collett. The writings of Agassiz, Cope and other naturalists have been consulted in the earnest desire to obtain a correct idea of our batrachian and reptilian fauna. Nevertheless, I have at times doubtless fallen into error.

The following is a summary of the species of each group known to occur in Indiana:

Tailed Batrachians	18 species.	
Tailless Batrachians	12 species.	
Total Batrachians		30 species.
Snakes	28 species.	
Lizards	5 species.	
Turtles	18 species.	
Total Reptiles		51 species.
Total of both classes		81 species.

THE AUTHOR.

IRVINGTON, IND., Sept. 1, 1892.

THE BATRACHIANS AND THE REPTILES OF INDIANA.

On the part of people who have not made a scientific study of animals no distinction is made between the group of creatures here called *Batrachians* and that group called *Reptiles*. The amphiuma and the snakes, the salamanders and the lizards, the common toad and the turtles are all called "reptiles." Nor is this strange when we consider how closely members of both groups resemble one another in outward form and in habits. It is indeed only recently that zoölogists, who endeavor to found their systems on more important differences than appear on the outside, have agreed to regard the frogs, salamanders, and newts, as fundamentally different from the lizards, turtles, and snakes. In reality, the batrachians are more closely related to the fishes than to the reptiles, while the latter are more nearly akin to the birds. The batrachians form a class standing intermediate between the class of fishes and the class of reptiles.

Nevertheless, since zoölogists have almost universally associated the two classes in their works, and since people do not usually distinguish the one kind of animals from the other, they are here described together.

The batrachians differ from the reptiles in several important respects. The skin of the former is usually smooth and moist, sometimes raised up into warts, as in the toads, but never disposed in overlapping scales or regular plates. Scales and plates, such as are seen in the lizards and snakes, and tortoises, are almost universal among the reptiles. No Indiana reptile is without such a covering, except our soft-shelled turtles. The life-history of the members of the two groups is also widely different. The batrachians almost always lay their eggs in the water, and the young pass their early days there as tadpoles. They respire by means of gills until the time of their metamorphosis approaches, when lungs are developed, the gills are absorbed, and the animal leaves the water and lives to a greater or less extent on the land. Reptiles, on the contrary, lay their eggs on land, the young are hatched with the form of the adults, and they never have gills. A few batrachians retain their gills life-long, breathing both by means of these and their lungs. Other differences exist, but since their determination would require dissections, they are not thought suitable for consideration in a work of this kind.

Since the animals herein described are a source of discomfort and alarm to many people, it may be well to say here that of all the batrachians and reptiles known to inhabit Indiana, but four, the yellow-banded rattlesnake, the prairie rattlesnake, the coral snake, and the copperhead, are poisonous. It is possible that the poisonous southern moccasin, or cottonmouth, may yet be found in the southwestern part of the State; if so, we shall have five poisonous species, and five only.

KEY TO THE CLASSES.

A. Skin usually smooth and soft, sometimes rough and warty, never forming scales that overlap or are arranged in regular rows; eggs usually laid in the water and giving origin to tadpoles. (Water-dogs, salamanders, frogs and toads.)
Batrachia, p. 5.

AA. Skin usually having epidermal scales or large regular plates; these usually arranged in a regular manner, often overlapping. Eggs laid on land. Young with form of adults. (Snakes, lizards, turtles, alligators, etc.) *Reptilia*, p. 73.

BATRACHIA.

The Batrachia include a great variety of animals that are found living in all except the coldest parts of the earth and the salt water. As already stated, they are, with rare exceptions, hatched in the water, where they spend at least a portion of their lives. A few forms retain their gills throughout life, and seldom or never leave the water. In a few cases the eggs are laid on the land, under sticks and stones; the young from such eggs may have very rudimentary gills and consequently never enter the water. Such species closely approach, in their habits, the reptiles. The gills may be either internal or external; usually they are of the latter kind. The external gills are attached to processes of the skin, and not to the branchial arches. The internal gills of the tadpoles or frogs grow out from the branchial arches, as in fishes.

The skin of the batrachians is richly provided with glands. These secrete a milky fluid, which is often acrid, and sometimes poisonous to the enemies of the species producing it. It thus serves as a means of defense to these animals, which are otherwise almost helpless. Often the glands are collected into groups, as in the case of those on the back of the head of the common toad. In some species the skin forms a fin on the upper and lower sides of the tail; but in such fins there are no rays, such as are found in the fins of fishes.

When limbs are present they have the same skeletal elements as the limbs of reptiles and mammals. Some batrachians are devoid of limbs. All of our species have the anterior limbs present; most of them have also the posterior pair. The anterior limbs never have more than four fingers; the posterior may have five toes.

Not much can be said here regarding the skeleton. The vertebræ are usually either amphicœlous or opisthocœlous. Ribs are often absent; when present they do not connect with a sternum below. In the lower forms as many as four branchial arches may be present; in the higher species the number is reduced. There may be teeth on the maxillaries, premaxillaries, vomers, and dentaries; more rarely on the palatines, the pterygoids and the splenials. A band of teeth may be found in some cases supported by the parasphenoid. The teeth are almost always very simple in structure, pointed, and grown fast to the supporting bones.

Breathing is effected in the adult by drawing the air into the mouth through the nostrils, then closing these, contracting the cavity of the mouth, and thus forcing the air into the lungs. Hence, a frog may be suffocated by holding its mouth open.

For additional information on the anatomy of the Batrachia the student should consult Prof. Huxley's article, "*Amphibia*," in the Encyclopedia Britanica; also for the Urodela, Dr. R. Wiedersheim's work, "*Kopfskelet der Urodelen.*"

The living species of Batrachia have been divided by Prof. E. D. Cope (*51*, 13) into four orders, viz: Proteida (*Necturus*), Urodela, Trachystomata (*Siren*), and Salientia. I prefer here to retain the genera *Necturus*, *Proteus* and *Siren* under the Urodela.

KEY TO THE ORDERS OF *Batrachia*

A. Limbs present or absent; when present, the hinder pair not much more strongly developed than the anterior. Tail developed or not; present in all our species. Animals fitted for creeping on or burrowing in the earth or for swimming in the water.
Urodela, p. 6.

AA All four limbs present and the hinder pair greatly developed. Tail wholly absent in the adult   Animal, when on land, usually progressing by leaping.   Salientia, p. 48.

Order URODELA.

Batrachia having a lizard-like, eel-like, or serpent-like form. All limbs, as well as the supporting girdles, absent in the extralimital *Cæciliidæ*. At least the fore limbs and the shoulder girdle present in all our forms; and usually also the hinder limbs. Posterior limbs never conspicuously larger than the anterior. Proximal elements of the tarsus not elongated. Vertebræ numerous, at least 14 in front of the sacrum; these either amphicœlous or opisthocœlous Ribs present, short. Maxilla present in all except *Necturus* and *Siren*. Teeth present on maxillaries, vomero-palatines, and on the dentaries, except in *Siren*. No tympanic cavities or eustachian tubes. Cloaca opening externally by a longitudinal slit.

The Urodela include about 133 species, distributed principally north of the equator. North America furnishes 54 species, 18 of which, at least, are found in Indiana.

The order as here defined contains 10 families. Of these, two, the Cœcilidæ and the Thoriidæ, are not natives of North America.

Key to the N. A. Families of *Urodela*.

A. Maxillary bone wanting. External gills present at all times of life.
    a. Body eel-like. No posterior limbs.   *Sirenidæ*, p. 8.
    aa. Body lizard-lke. Two pairs of limbs.   *Proteidæ*, p. 10.

AA. Maxillary bone present. No gills in the adult state. All four limbs present.
    a. Body extremely elongated. Both pairs of limbs very rudimentary.   *Amphiumidæ*, p. 12.
    aa. Body lizard-like. Anterior and posterior limbs well developed.
        b. A branchial slit on each side of the neck. Vomero-palatine teeth close to and parallel with those of the premaxillaries and maxillaries.
            Cryptobranchidæ, p. 15.
        bb. No branchial slits on the side of the neck. Vomero-palatine teeth considerably behind the premaxillary teeth.
            c. Parasphenoidal teeth present, forming two brush-like bands along the roof of the mouth.
                Plethodontidæ, p. 31.
            cc. No parasphenoidal teeth.
                e. Vomero-palatine teeth in a transverse row between, or just behind, the choanæ.
                    Ambystomatidæ, p. 17.
                ee. Vomero-palatines in two longitudinal rows along roof of the mouth, on two prolongations backward of the vomers.
                    Salamandridæ, p. 44.

Family I. SIRENIDÆ.

Body elongated, eel-like. Posterior limbs wanting; the anterior moderately developed, with three or four fingers. External gills present during whole of life. One to three branchial slits on each side of the neck. Premaxillaries and dentaries without teeth, and covered with a horny sheath. Teeth on the splenials. Vomers provided with teeth, which form two converging patches on the roof of the mouth. No maxillaries present. Contains two genera, each embracing a single species, both confined to North America.

KEY TO THE GENERA OF *Sirenidæ*.

A. Fingers four; branchial slits usually three.   *Siren*, p. 8.
AA. Fingers three; branchial slit one.   *Pseudobranchus*.

Pseudobranchus has not been found north of Georgia and Florida.

Genus SIREN, Linnæus.

Siren, Linnæus, 1765, *95*; i, 311; Boulenger, 1882, *28*, 86; Cope, 1889, *51*, 225.

Hand with four fingers. Jaws with horny sheath. Vomerine patches of teeth broad. Tongue broad, free in front. External gills and branchial slits, three. Eyes distinct, but without lids.

Siren lacertina, Linn.

Mud-eel. Siren.

Siren lacertina, Linnæus, 1765, *95*, i, 311; Holbrook, 1842, *54*, v. 101, pl. 34; Boulenger, G. A., 1882, *28*, 87; Cope, E. D., 1889, *51*, 226; Barton, B. S., 1821, *40*.

Body long, slender, nearly cylindrical; about twelve times as long as the head is broad. Head flat above, sides tapering to the blunt snout, which overhangs the mouth. Gill slits three. External gills three pairs; but these appear to be abortive in the young up to six inches in length. Eyes small, without eyelids. Premaxillaries distinct, toothless, covered with a black horny sheath. Vomers and palatines distinct, each with several rows of small teeth. Dentaries also toothless, and covered with a black horny sharp-edged sheath. Teeth on splenials (just below hinder ends of dentary sheath). Tongue free on sides and in front.

Anterior limbs well developed, short, furnished with four digits each. No trace of hinder limbs. Body and tail crossed by numerous furrows; about thirty of these from fore legs to vent. Tail compressed, with a fin above and below; its length somewhat more than one-third the total length, terminating in a sharp point.

Color a bluish black, paler below. Sides of head sometimes with a yellow band from muzzle to gills.

The length attained may be as great as thirty to forty inches.

Distribution from North Carolina west to Mexico, and up the Mississippi Valley to Alton, Ill., and Lafayette, Ind.

Indiana localities: New Harmony (Sampson's coll.); Mt. Carmel, (Nat. Mus. coll.); Terre Haute (B. W. Evermann); Lafayette (S. Coulter); Washington, Daviess County (C. H. Gilbert).

HABITS.—One of the earliest and best accounts that we have of the habits of this curious animal was given by Dr. B. S. Barton, of Philadelphia, in a letter addressed to John Gottlob Schneider, the author of "Historia Amphibiorum." Schneider and some other authors supposed that the *Siren* was only the larva of some yet undiscovered salamander. Barton argues against this notion with ability. He states that the favorite resorts of the *Siren* are the rice ponds and the adjacent muddy grounds. It sometimes attains a length of 20, 30, or even 40 inches. Where it abounds it is, according to Barton, called "Alligator" and "Water-lizard." He kept one for almost a year, and experimented with it. During this time, it underwent little or no change in either size or condition of its branchiæ. While in the water, the gills are incessantly in motion. The animal appears to depend only partially on its gills for oxygen, since various observers have noted its coming frequently to the surface for air. Barton says that in warm weather it comes to the surface every five minutes, while in colder weather it comes up about every fifteen minutes. It has the power to remain under water for more than an hour at least. It appears to spend a good portion of its time burrowing in the mud of swamps. Barton thinks that it sometimes travels overland, although its movements on land are slow. An earlier observer, Garden, who corresponded with Linnæus, declares that the *Siren* may crawl up on the trunks and limbs of fallen trees, likewise that it sings with a querulous voice similar to that of a young duck. This idea gave origin to the name *Siren*. It possibly emits a shrill, somewhat hissing tone like that of the Amphiuma. Barton appears to think that the *Siren* can take water into the mouth through the nostrils; an interesting observation in view of the fact that some salamanders do the same as a means of breathing. Cope (*51*, 224) has recorded his observation that in small specimens the branchiæ are rudimentary and apparently functionless, and that it is only in the adult individuals that the gills are fully developed.

Linnæus affirms that the *Siren* lives on serpents which it catches and holds with its strong teeth. Barton doubts this, but considering the great voracity of the Batrachia in general, and the large size reached by some of the *Sirens*, it appears quite reasonable. Their teeth, however, are not "strong." Barton fed his specimen on angleworms, pieces of meat, etc. It takes its food with "inconceivable rapidity." It can endure hunger for long periods. The one kept by Barton was exposed for several days in water at a temperature of 33° or 34° F, and for a few hours was "locked up, as it were, in the ice."

Holbrook (*54*, v. 101) says that this animal lives chiefly in the mud, but this they sometimes leave and take to the water, in which they swim with great swiftness. He says too that they are sometimes found on the land, but he did not know why they go there.

So far as I am aware, nothing is known concerning the breeding habits of this species. It is so rare in our State that few opportunities are likely to be offered to any observer to study such habits. This will have to be done in the South, where it abounds. I have not seen mention made of specimens having been found of less than three inches in length. Specimens of this size are very desirable.

Family II. PROTEIDÆ.

Body lizard-like in form; limbs, four; external gills, three pairs, present during whole of life; maxillaries, absent; teeth on premaxillaries, vomers, palato-pterygoids, and dentaries.

Only the genus *Necturus* is found in North America. *Proteus* is an inhabitant of the subterranean waters of Austria.

Genus NECTURUS, Rafinesque.

Rafinesque, 1819, *97*, 417; Boulenger, 1882, *28*, 84; Cope, 1889, *51*, 23.

Fingers 4; toes 4; permanent branchial slits 2 pairs; eyes exposed; tongue large; vomero-palatine teeth strong, in a single row; tail broad and compressed.

Premaxillary teeth 11–15; vomerines 12–16; spotted.
maculatus, p. 10.

Premaxillary teeth 6–8; vomerines 8–9; no spots. N. C., S. C.
punctatus.

Necturus maculatus, Raf.

Water-Dog; Gilled Salamander.

Necturus maculatus, Rafinesque, 1819, *97*, 417; Boulenger, 1892, *28*, 84; Cope, E. D., 1889, *51*, 23, with figures.

Triton lateralis, Say, 1823, *14*, 5.

Menobranchus lateralis, Harlan, 1825, *62*, i, 221; Holbrook, 1842, *54*, 115, pl. 38.

Necturus lateralis, Wagler, 1830, *75*, 210.

Body moderately elongated, thick, cylindrical or moderately depressed, and with a distinct dorsal groove. Head broad and flat. Snout rounded. Labial folds well developed. A distinct gular fold. Eyes small. The two series of teeth in the upper jaw near to, and parallel with, one another. Premaxillary series short; the vomero-palatine series extending back to the corner of the mouth. The mouth is large and extends backward to under the eyes. Nostrils in the edge of the lips. Head contained in the distance to the vent about five times; three-fourths as wide as long. Costal folds usually 14. Tail broad and compressed, one-third the total length. Limbs well developed. Digits 4-4.

Skin smooth and furnished with numerous follicles.

Color ashy brown, palest below, and more or less mottled and spotted with blackish, especially above. Young specimens with a lateral dark band from the snout to the tail. Gills bushy, and in life bright red.

Prof. Cope (51, 26) mentions specimens from Ontario which were entirely black.

This species reaches a length of 24 inches or more.

This is a widely distributed animal in Eastern North America, and in places it is very abundant. It is recorded from the territory extending from Montreal to Alabama, west to Wisconsin, Kansas, W. Arkansas, and Louisiana.

It will no doubt be found in all parts of Indiana, but it is very common in the numerous lakes of the northern portion of the State. I have record of it from Lafayette (S. Coulter); Mt. Carmel, Ill. (National Museum); New Harmony (Sampson's coll.); Monroe County (Ind. Unv. coll.); Franklin County (Hughes). I have found it common at Lake Maxinkuckee, Marshall County. Prof. Blatchley reports it as very common at Terre Haute. Of 19 specimens sent me by Prof Evermann more than half had the gills very short or almost wholly gone.

HABITS.—This species appears to be wholly aquatic, although it does not depend on its gills entirely for the oxygen that it consumes. It has been observed to come to the surface for air; and its gills are sometimes missing, as though nibbled off. In such cases breathing must be accomplished by other means. They have quite well developed lungs, which some observers have artificially inflated. The animal has seldom been observed to go on land. (54, loc. cit.) In the water they progress either by creeping along on the bottom or by swimming. The swimming is accomplished mainly by strokes of the large flat tail, and their progress is rapid. They are, for the most part, nocturnal in their habits.

The food of the Water dog consists of insects, crustaceans, worms and mollusks. They are frequently taken on the hooks of fishermen. They will, no doubt, be found to be willing to eat almost anything that can serve as food. They are not acceptable game to the fishermen, since the latter regard them as very poisonous, and dislike to handle the creatures. This notion is an erroneous one; but a nip by their sharp and strong teeth would no doubt be somewhat painful. Drs. Wilder and Barnard (22, 8,

438) cooked and ate one, and reported it as excellent. The animal might, therefore, possibly be made useful, if people could overcome their prejudices.

The Water-dog has the power of enduring long periods of cold and hunger. Kneeland reports having specimens for three months in water that froze more or less every night.

Dr. C. O. Whitman states that the egg of *Necturus* is about the size of a pea, and, unlike the eggs of most batrachians, it has no pigment to obscure the processes of development. It forms an admirable object for the study of batrachian development.

Large numbers of the *Necturus* are found in the streams emptying into Lake Michigan and along the shores of the lake. At Ecorse, Michigan, 2,000 were taken in a minnow seine at one haul. It is stated that some of these were so gorged with white fish spawn that when they were thrown on shore the spawn flew out of their mouths. Another man, according to Professor J. W. Milner (Report of the United States Fish Commission, 1872–3, p. 62), had out, at Evanston, Illinois, 900 fish-hooks, and from these, in one day, he took 500 of the "lizards." Milner states that they make no more attempt to bite than does a frog. A full series was collected from the Detroit River of specimens ranging from an inch and a half to full grown. About the middle of July eggs were taken.

Family III. AMPHIUMIDÆ.

Body elongated and eel-like. Limbs two pairs, feebly developed. No external gills in adult stage. Tail well developed. Skull long and narrow. Maxillaries large, directed backward, toothed. Premaxillaries united. Vomero-palatines narrow, directed backward nearly parallel with axis of head. No true ethmoid.

A family containing a single genus.

Genus AMPHIUMA, Garden.

Amphiuma, Garden, 1821, *98*, i, 599; Boulenger, 1882, *28*, 82; Cope, 1889, *51*, 215.

Limbs very feebly developed; digits two or three on each. A single branchial slit on each side the neck. Teeth of maxillary and vomero-palatine in two parallel, backwardly directed series along each upper jaw. Premaxillaries united; developed from a single center; sending back two strong processes, one in roof of mouth, the other between nasals and frontals on the upper surface of the snout.

Contains but a single species, confined to North America.

Amphiuma Means, Garden.

Amphiuma; Congo Snake.

Garden, *loc. cit.* Holbrook, 1842, *54*, 5, 89, pl. 30; Boulenger, 1882, *28*, 83; Cope, 1889, *51*, 216, with figures.

Body long and eel-like. Head narrower and more pointed than is usual among the batrachians. A single gill-slit on each side. Eyes extremely small, barely seen through the skin. Maxillary and vomero-palatine teeth forming four nearly parallel rows in the roof of the mouth. No external gills in the adults. Fore and hind limbs present, extremely feeble in development. Digits on each somewhat variable, usually two or three. Length of head (snout to gill-cleft), in length from snout to vent about ten times. Tail about one-fourth the total length, compressed, slender and pointed. Skin everywhere smooth.

Color dark slaty or reddish brown, paler below. Lower jaw and edge of upper lip yellowish.

The amphiuma may reach a length of three feet. It is found from the Carolinas west to Louisiana. The author has taken it at Little Rock, Arkansas, and has seen a specimen in the National Museum at Washington which was taken at Jeffersonville, Indiana, by Mr. George Spangler. The specimen is fifteen inches long and was received at the Museum March 25, 1880. Careful observations along the Ohio and Wabash Rivers will no doubt result in bringing additional specimens of this interesting animal to light.

This, like the Siren, appears to be a mud-loving species. Its whole structure appears to adapt it to burrowing about in the mud at the bottoms of creeks and rivers and ditches. Its head is long and pointed, and the bones of the skull firmly bound together, as if to render the head the point of a drill. The first instinct of the animal, when put into any vessel, seems to be to burrow out of sight. This habit of burrowing in mud has been observed ever since the discovery of the creature. Harlan (*39*, 86) speaks of them as "burrowing in the mud in swamps, or in the vicinity of streams, where it searches for its food and hibernates, occasionally visiting the dry land." Other specimens are spoken of as having been found several feet beneath the recent alluvial deposit, under the decayed trunk of a tree. The same author (*39*, 188) states that he had been informed that "they are sometimes discovered two or three feet under mud of the consistency of mortar, in which they burrow like worms, as was instanced in digging near a street in Pensacola, when great numbers were thrown up during the winter season."

The food of this animal consists of a variety of aquatic animals. Harlan says that in the stomach of some were found small fishes and beetles. Holbrook adds to their diet small mollusks.

Both Harlan and Holbrook note the habit possessed by this animal of leaving the water occasionally, but the purpose of so doing was not known. No doubt it passes from one water to another in this way, since it can creep readily and does not depend on the water for respiration.

The writer has had the opportunity of studying the breeding habits of the Amphiuma. At Little Rock, Ark., on the first of September, a female was found lying in a small excavation underneath a fallen tree trunk in a cypress swamp. The tree was at a considerable distance from any water. Her body was disposed in a coil, in the midst of which was a mass of eggs. In these eggs were young so far advanced that they would soon doubtless have been excluded. The young, which constituted the whole contents of the eggs, were surrounded by a transparent capsule about as thick as writing paper. The eggs were a little more than a third of an inch in diameter, and were connected by a cord of substance similar to that of the capsule itself. This cord varied in length from a fifth to a half inch. There appeared to be two strings in the mass, but of this I was not certain. The whole mass of eggs resembled closely a string of large beads. Of the eggs there were probably about 150.

Within the eggs the young were coiled in a spiral. Their length is something less than two inches. Each had three pairs of conspicuous gills, and, since they were evidently nearly ready to hatch, it is probable that they would retain the gills for awhile after entering the water. The gills consisted of a main stem, from which were given off eight or ten branches. Three gill-slits remained open. The eyes appeared to better advantage than they do in the adults. Though the gill slits are probably present in the just hatched animal, they certainly do not remain long, since Harlan mentions (*39*, 189) having received specimens of only three inches in length that did not exhibit the least appearance of gills.

Not only does the amphiuma exhibit to a remarkable degree, for so low a creature, the maternal instinct, but it is an animal of a good measure of spirit. Dr. Shufeldt (*50*, 2, 163,) experimented somewhat with the living animal. Into the vessel containing the amphiuma he threw a dead snake. This the amphiuma seized quickly with its jaws and extending itself, began to whirl around rapidly in the water. After the snake had been released, it was again thrown to the amphiuma, and the movements were repeated. When the dead ophidian was presented to the amphiuma the third time no attention was given it. While the female that I captured was still alive, I placed her on a floor and poked her a little with a stick. This she seized, and springing from the floor, she whirled round and round in a spiral form and turned the stick in my hand unless I held it tightly.

These animals are remarkable for the size of their blood corpuscles, the largest furnished by any vertebrated animal, since they can be seen by the naked eye. The number of their vertebræ is also worthy of remark, there being altogether about 110, of which sixty-five lie in front of the sacrum.

Family IV. CRYPTOBRANCHIDÆ.

Form salamandrine. Head broad and depressed, mouth wide. Limbs four, well developed. Tail broad and compressed. Teeth on the anterior edge of vomers, concentric with those of premaxillaries and maxillaries, but not extending so far back as latter. No teeth on parasphenoids.

Contains two living genera, *Megalobatrachus* of Japan, three feet long, and the following:

Genus CRYPTOBRANCHUS, Leuckart.

Cryptobranchus, Leuckart, 1821, *77*, 260; Boulenger, 1882, *28*, 81; Cope, 1889, *51*, 37; *Menopoma*, Harlan, 1825, *62*, 221.

Tongue large, free in front. One gill-slit on each side. Four branchial arches. Maxillary row of teeth extending back to angle of mouth; the vomerine to choanæ; the two series lying close together.

Tongue free in front; its upper surface with folds.
alleghaniensis, p. 15.
Tongue not free; its upper surface with large papillæ. Tenn.
fuscus.

Cryptobranchus alleghaniensis, (Daudin.)

Alleghany Salamander; Hellbender.

Salamandra alleghaniensis, Daudin, 1803, *69*, viii. 231; *Cryp. salamandroides*, Leuck., l. c.; *Menopoma alleg.*, Harlan, l. c., with figures; Holbrook, 1842, *54*, v. 95, pl. 32; *Cryp. alleg.*, Van der Hoeven, *104*, iv. 384; Boulenger, *28*, 81; Cope, 1889, *51*, 38, with figures.

Size large, body heavy and depressed. Head broad and flat and snout rounded. Tail broad and much compressed, and with a fin along its upper edge; its length equal to half the remainder of the animal. Skin richly provided with mucous follicles, especially about the head. Sides with a conspicuous, corrugated cutaneous fold, which extends from the angle of the mouth to the middle of the tail.

Mouth large, the gape extending to behind the eyes. Nostrils small, close to the edge of the lips. Eyes small, with no evident lids. Limbs short and stout, bordered externally by a fold of membrane, which extends down to the outer toe. Hinder limb stoutest, and bordered behind by a second fold, which, near the foot, passes into the external fold. Digits 4–5.

The distance from the snout to the gill-slit is equal to about one-sixth the total length.

Color grayish to black, usually dark slate; sometimes mottled and spotted.

A total length of two feet may be reached, though usually smaller.

Known to occur from Pennsylvania west to Iowa, south to South Carolina and Louisiana. Not yet known from Florida or Texas.

Indiana localities are: Evansville (Nat. Mus.); Whitewater R. at Brookville (E. R. Quick); Ohio R. at Vevay and Madison (Hay.); New Harmony (Max. von Wied, *103*, xxii. 136).

It appears strange that the occurrence of this animal has not yet been reported from any part of the State except immediately along the Ohio River, since its extra limital distribution indicates plainly that it must be found in all the larger lakes and streams of the State.

The "Hellbender" is a great coarse, sprawling, repulsive looking creature. In the Ohio and its tributaries it is a common animal, and is cordially despised by the fishermen, whose hooks are too often burdened with it instead of fish, and who regard it as also poisonous. The latter idea is erroneous; nevertheless, its bite might be painful. They appear to be exclusively aquatic, and yet are capable of living for a long time out of the water. Mr. Charles H. Townsend (*22*, 16, 139), says that in Loyalhana Creek, Westmoreland County, Pa., shoals of them may be seen in early spring moving sluggishly over the bottom. In August they appear to hide away under rocks, but will come out after a bait of dead fish. They are very tenacious of life. Mr. Townsend carried some of them in a bag, on horseback, for five miles through a blazing sun, then put them into a tub of water for five weeks without food, and they survived and were vigorous. Mr. Wm. Frear, of Lewisburg University, in corroborating Mr. Townsend's account of their tenacity of life, states that a specimen eighteen inches long had lain in the sun forty-eight hours, then in a museum a day longer, before it was placed in alcohol. After being submerged in this twenty-four hours it was taken out and was alive and vigorous. On making cuts, in the attempt to sever the cervical vertebræ, it showed no anger or discomfort; but if only the tip of the tail was touched, the animal would viciously snap its jaws and strike with its tail.

Two females of Mr. Townsend's specimens spawned while in the tub where he put them. The spawn is reported as similar to that of frogs, but without the dark colors of the latter. The ova were exuded in strings, and were farther apart than the eggs of frogs. The eggs were of a yellowish color and surrounded with gelatinous matter, which swelled greatly on coming in contact with the water. Professor Cope states that the eggs are rather large and are attached by two strong suspensors at opposite poles. They must resemble somewhat the eggs of the Amphiuma, but with soft gelatinous covering.

The very young have not yet been seen, so that it is not known whether or not they possess gills. If gills are present on hatching, they must be very transient, at most.

Maximillian, Prinz von Wied (*103* xxii, 136), states that he found the young of this species at New Harmony. They were about three inches long, had two gill-slits yet open, and remains of external gills. These gills were afterward gradually absorbed. It is possible, however, that these were the young of some *Ambystoma*, as Professor Cope has suggested to me.

The Hellbender is said to eat worms, crustaceans and fishes. It will probably eat almost anything that presents itself.

Grote states that he has seen these animals with the shed skin rolled up in their mouths, as if in the act of swallowing it. They were observed during July and August to have a swaying motion of the body, but the purpose of this was not discovered. The statement is also made that during the time of pairing of the sexes the tail becomes broader and the lateral folds undergo some changes. The eggs seem to be deposited in the water during the month of August.

Professor S. P. Gage and wife have made the observation that this animal at times draws in water by pharyngeal movements and expels it through the gill-slits.

Family IV. AMBYSTOMATIDÆ.

No persistent gills or gill-slits. Limbs four; digits four to five. Vomers broad, with the teeth on the posterior edge, the vomero-palatine teeth forming a nearly straight row across the roof of the mouth, in a line just behind the choanæ. No parasphenoidal teeth. Tongue large, with a narrow anterior and lateral free edge. Carpus and tarsus ossified.

Of the *Ambystomatidæ*, as here defined, all the species belong to North America, except a single one, and that occurs in Siam. Of the N. A. forms Professor Cope recognizes four genera. Since these genera are founded on peculiarities of structure of the hyoid apparatus, and these require minute dissections, and since all known Indiana species belong to the genus *Ambystoma*, the others are not defined. The genera recognized by Professor Cope are *Amblystoma*, *Chondrotus*, *Linguœlapsus* and *Dicamptodon*.

Genus AMBYSTOMA, Tschudi.

Figure 1, *Plate* 1.

Ambystoma, Tschudi, 1838, *29*, 92; *Amblystoma*, Agassiz, 1848, *100*. Larval forms.
Siredon, Wagler, 1830, *75*, 209.

Axolotes, Owen, 1844, *101*, xiv, 23.

Vomero-palatine teeth in a nearly straight, sometimes interrupted, row behind the choanæ. Tongue broad, free margin narrow, the upper surface with numerous narrow folds. Tail usually somewhat compressed.

As formerly defined, this genus would comprise about eighteen species. As defined by Professor Cope, it will contain still about thirteen species. Of these we certainly have in Indiana six, and possibly more.

The Ambystomas are familiarly known as "Mud-puppies," and are frequently met with in the operations of plowing, digging ditches and cleaning out cellars. They are often not distinguished from lizards, but the latter are very active, while the former are extremely slow in most of their movements. The Ambystomas, like all similar animals, are regarded as venomous and are usually destroyed on sight. On the contrary, they are the most innocent animals imaginable. They can not be provoked to bite; and if they should do so, they have no poison glands connected with their teeth, and the teeth are so very short that they could scarcely penetrate the skin.

The species appear to be mostly nocturnal in their habits. Their food in the adult stage consists of insects, small mollusks, and more especially of earthworms. Their winters seem to be spent near the borders of ponds, and in these, as soon as the ice is broken up in the spring, their eggs are deposited. These are surrounded by a mass of gelatine, by means of which they cling to one another and to grass and sticks. The tadpoles are slenderer than are those of toads and frogs, and they have branching external gills. They look a good deal like small catfishes In due time these tadpoles lose their gills and betake themselves to the land. The larvæ of some of the species attain a large size before undergoing their metamorphosis, and one species has been known to breed while still possessing gills.

KEY TO THE N. A. SPECIES OF *Ambystoma*.*

A. Tongue with a median crease from which the papillary folds diverge in a pinnate manner. Palatine teeth not extending beyond the choanæ. *microstomum*, p. 19.
AA. Tongue with the papillary folds radiating from a point at the back of the tongue. Palatine teeth extending laterally beyond the choanæ.

---

*In determining the species of Ambystoma great care should be exercised. Too much reliance must not be placed on this key, but the descriptions must be appealed to, and the animal in hand carefully compared with them. Those costal grooves are to be counted which plainly lie between the fore and the hind legs. The number of these, in any species, may occasionally be one more or less than here given. Examine both sides of the body. The color and character of the spots may vary within limits.

a. Costal grooves 12.
  b. A single indistinct plantar tubercle or none. Black on sides, with a yellow stripe along the back. Head broad. Idaho. *epixanthum*.
  Leaden brown, with a grayish-brown dorsal stripe. Wyoming to Pacific coast. *macrodactylum*.
  Lead color to black, sometimes with whitish specks along the sides. *jeffersonianum*, p. 22.
  bb. Two distinct plantar tubercles.
    Lower jaw projecting beyond the upper.
        xiphias. Appendix.
    Lower jaw not projecting perceptibly beyond the upper. Sides with vertical yellow spots. N. Mexico. *trisruptum*.
    Lower jaw not projecting. Spotted irregularly with yellow. *tigrinum*, p. 23.
aa. Costal grooves 11.
  c. Two distinct plantar tubercles.
    Dark brown above, brownish yellow below. Tail equal to rest of body. *copeianum*, p 26.
    Dark above, with large, irregular yellow spots. Yellow below. *bicolor*.
  cc. No plantar tubercle, or one indistinct.
    Lead color. Tail equal to distance from vent to gular fold. Size small. Pennsylvania to Georgia.
        conspersum.
    Dark above, with a row of yellow spots of size of eye on each side of back. *punctatum*, p. 27.
    Black above, with cross bands of gray. *opacum*, p. 29.
aaa. Costal grooves 10.
  Brown, with sprinklings of gray. *talpoideum*. Appendix.

Ambystoma microstomum, (Cope.)*

Small-Mouthed Salamander.

Figure 1, *Pl.* 1.

Amblystoma microstomum, Cope, 1867, *1*, 206; Boulenger, 1882, *28*, 50; *Chondrotus microstomus*, Cope, 1887, *22*, 88; 1889, *51*, 101, with figures.

Body longest and slenderest of all the species of *Ambystoma*, the distance from the snout to the axilla being contained twice in the distance from the axilla to the groin. Head small and short; contained in length

---

*Although Prof. Cope has assigned this species to his recently proposed genus *Chondrotus*, I retain it in *Ambystoma*; since a careful examination of the tongue structures has shown that there is no essential difference between these and those of other species of *Ambystoma*. The otoglossal is identical with that of *A. opacum*.

to the groin 6 or 7 times. Snout rounded. Lower jaw slightly projecting beyond the upper. Mouth small. Tongue of moderate size, oval, the papillary folds arranged pinnately on each side of a conspicuous median groove. Palatine teeth extending little, if any, beyond the choanæ; forming a nearly straight line across the palate, or with an obtuse angle pointing forward.

Body with 14 costal grooves, and a slight dorsal furrow. Tail a little longer than the distance from the fore to the hind limbs; nearly cylindrical at the base, becoming flattened toward the obtuse tip. Limbs short and weak, failing much of meeting when laid along the side.

Skin smooth and slippery, and with numerous minute mucous pores.

The color above is a dark brown or black, paler below. In alcohol there is sometimes a tinge of red. All over the body, but especially along the sides, there are numerous mildew-like specks of gray.

The extreme length may become as much as six inches.

This species has been found inhabiting the country from Ohio to South Carolina, and west to E. Kansas and Louisiana. It has also been brought from about Hudson Bay.

In some parts of Indiana it is an extremely abundant species. It is the commonest species of Urodele about Irvington. I saw it in Mr. Sampson's collection at New Harmony. Dr. Ridgway collected it in numbers at Wheatland. It has been taken at Terre Haute by Profs. Evermann and Blatchley, and at Brookville by Mr. A. W. Butler. One specimen in the Terre Haute Normal School collection from Howard County.

I have been enabled to study this species with some care, and I shall detail its habits somewhat, since they will illustrate the habits of other species of the genus.

The Small-mouthed Salamander spends the winter months in and about the borders of stagnant ponds. I have had it taken in midwinter from under ice over which boys were skating. It was captured under such circumstances on January 6, 1886, and again February 6, 1889. These winters were rather mild, and the ice not thick. As soon as the ice is melted they begin to lay their eggs, and it is no unusual thing to have the ponds frozen solidly again after some eggs have been deposited. Some specimens in captivity laid their eggs in the middle of January. Oviposition more commonly occurs during the month of March. It may continue for a space of at least three weeks.

The eggs proper are quite small, being about one-twelfth of an inch in diameter. Each egg is surrounded with a capsule of gelatine, which on coming in contact with the water swells up and becomes about as large as a hazel-nut. By means of this gelatinous mass the eggs are cemented to one another and to leaves of grass, sticks, and other objects in the water. The eggs may be deposited singly, but more commonly they occur in masses of from two to a dozen, and even more. This gelatine no doubt serves to protect the eggs and the young. I do not believe that any animal will eat the gelatine.

Segmentation of the eggs begins immediately, and soon the young salamander may be seen within the gelatine looking like a dark worm. The development of some of the eggs appears to progress more rapidly than that of others. The young escape into the water from the first of April to the tenth. They are then about two-fifths of an inch long, are of a bright olive color, with some squarish blotches along the back. Three little gills stand out on each side of the neck, and on these may be seen rudiments of lateral filaments. The fore-legs are seen as the merest rudiments. I do not think that at this stage they have any mouth. The body is covered with cilia. From the time of hatching there is a club-shaped organ attached just behind each corner of the mouth. These are the "balancers." By means of these the little animal manages to hang fast to the sides of the vessel, or to objects in the water. They are lost before the tadpole becomes an inch long.

When the larva has become a half inch long, its gills have developed two rows of filaments on each stalk. Through these the blood may be seen circulating. Water enters the mouth through the nostrils and passes out through the gill-slits, thus bathing the gills.

Larvæ two-thirds of an inch in length had their intestines filled with the remains of small animals. These consisted principally of entomostracous crustaceans. The stagnant waters where the salamanders lay their eggs swarm with minute animal life, and the tadpoles have sharp appetites and the ability to provide for themselves. Later on they seize on small mollusks and insects.

When they have become an inch long, they may be seen to come to the surface after air. This happens more and more frequently as the size increases. When they have attained a size of nearly two inches, the gills begin to show signs of undergoing absorption. They seek the surface oftener, and are inclined to float on the water. They will even leave the water if allowed to do so. The time of completing their change into the adult form is about the first of June. The gills are lost; the tail loses its fin-like border; the legs have become well developed; and soon the only difference between the animal and its parents is in size. They betake themselves to the banks, where they are soon concealed among grass roots and earth.

About the ponds the adults may often be found concealing themselves under sticks and pieces of wood which lie partly in and partly out of the water. Here they can get an abundance of their favorite food, the earthworms. After their eggs have been disposed of, they appear to leave the vicinity of the water, and scatter out, so that they are only occasionally found under logs and in the soil. At this period they do not seem to be able to remain a great length of time in the water, and when left there they have sometimes drowned.

I have often been surprised at the ability for climbing shown by these animals. I have seen them climb up the corners of a zinc box six inches high, and make their escape. Indeed, it is extremely difficult to keep them imprisoned.

The salamander will soon learn to take food when it is offered. They seize and swallow earthworms with great greediness. A worm several inches long will be swallowed by installments.

This species of *Ambystoma*, even in its adult stage, enjoys a sort of aquatic respiration. I have observed them, while under the water, to draw in streams of water through the nostrils and to expel it at intervals through the mouth. These expulsions of water by the mouth occur every eight to twelve seconds. The animal will remain under the water breathing thus for perhaps fifteen minutes. It will then appear to become uneasy, bubbles of air escape from its mouth, and soon it comes to the surface after air. In all probability oxygen is absorbed from the water by the lining of the mouth, and thus the blood is partially purified. In this way we may account for the long time that these animals can remain in water under the ice.

This salamander appears to enjoy having its back tickled with a straw. It will often lift its tail high in the air and wave it to and fro in a ludicrous way. It seems at times to make a low piping sound, and at other times produces a clucking noise.

It is of interest to us that this species was first described by Dr. E. Hallowell (*1*, 1856, 8) as *Ambystoma porphyriticum* from specimens sent him from Indiana.

Ambystoma jeffersonianum, (Green.)

Jefferson's Salamander.

Salamandra jeffersoniana, Green, 1827, *102*, 4; Holbrook, 1842, *54*, 5, 51, pl. 14; *Ambystoma jeffersoniana*, Baird, 1849, *2*, 283; *Amblystoma jeffersonianum*, Cope, 1867, *1*, 195; Boulenger, 1882, *28*, 46, pl. II, fig. 2; Cope, 1889, *51*, 101, with figures.

A species with an elongated body and head. The distance from the snout to the axilla considerably less than the distance from axilla to groin. Tail equal to distance from snout to vent. There are twelve costal grooves and a dorsal furrow. Skin smooth, but under the lens showing numerous mucous pores. A line of these internal to the orbits. Head rather broad and depressed; its width contained in length to groin from four to five times. Cleft of mouth extending back more than half way to gular fold. Eyes large and set far back. Gular fold prominent, but not meeting above. Tongue large and with the papillary folds radiating from behind. Choanæ apart same distance as the nostrils. Series of palatine teeth interrupted in the median line and behind each choana, thus making four patches.

Limbs long and slender; in some specimens overlapping when laid along the side; in others not quite meeting. Toes long and slender. Tail compressed, narrower above, but without a crest. The tarsal tubercles are lacking, or there may be one indistinct.

The body varies above from lead color to brown and black, paler below. On the sides of body and tail, sometimes also on belly, there may be distinct or faint blotches of bluish, but often these are wholly wanting.

A length of five to six inches may be reached by full grown adults.

Of this species Prof. Cope recognizes three varieties, or subspecies, viz.:

Head broad; its width in length to groin 4 to 4.5 times; lead colored to brown, sometimes with whitish spots. *jeffersonianum.*

Head (as above) 5 times; black, with numerous white spots on sides, belly and tail. *laterale.*

Head (as above) 5 times; uniform lead color. *platineum.*

A widely distributed species, but most common toward the north. Hudson's Bay south to Virginia and Southern Illinois. Indiana localities: Hanover (Hallowell); Irvington (Hay); Franklin County (A. W. Butler); Wabash County (A. B. Ulrey); Monroe County (Ind. Univ. coll.); Terre Haute (Blatchley); Lake Maxincuckee, Marshall County (the variety *laterale*). The latter variety with black skin, relieved by numerous bluish white spots on sides of body and tail and on the belly, and with narrow head, may be expected throughout northern Indiana. It may be distinguished from *Plethodon glutinosus* by the lack of parasphenoidal teeth.

The special habits of this species have not, so far as I know, been studied. These, however, we may be sure, resemble much those of the other *Ambystomas*; but we may be as certain that they will present some interesting peculiarities.

Ambystoma tigrinum, (Green.)

Tiger Salamander.

Salamandra tigrina, Green, 1825, *2*, v, 116; *Triton tigrinus*, Holbrook, 1842, *54*, v, 79, pl. 26; *Ambystoma tigrina*, Baird, 1849, *2*, 284; *Ambystoma mavortia*, Baird, 1849, *2*, 284; *Amblystoma tigrinum*, Cope, 4867, *1*, 179; 1889, *51*, 68, with figures; Boulenger, 1882, *28*, 43.

This species, as now recognized, is one of the most widely distributed of North American urodeles, being known from Maine to Florida, west to California, and southwest to Mexico. In this wide extent of territory the species has been subjected to a great variety of conditions, and the result has been the production of numerous forms, which differ so much that they have been described as distinct species under many names. The discovery of intermediate specimens has resulted in bringing all the forms under the earliest name, *tigrinum*. I describe the species as represented in Indiana.

A large, heavily built species, with long tail, stout limbs and a dark color, varied with numerous yellow spots.

Head about as wide as the body, or somewhat narrower in large individuals; the width in the length to the groin about four times. Parotoid region much swollen. Gular fold prominent, overlapping, rising on the sides of the neck. A groove running back from the orbit to the gular fold. Mouth large, reaching back half way or more to the gular fold. Tongue large, with the papillary folds radiating from behind. Vomero-palatine teeth in a straight or slightly curving series across the roof of the mouth, just behind the choanæ; occasionally the series projects forward in an angle between the choanæ.

The body is usually thick and depressed, and looks as if swollen. The distance from the snout to the axilla is contained in the distance from snout to groin two and a half times. There are twelve well marked costal furrows. The tail is commonly equal to the distance from snout to the groin, but specimens are often found in which it is considerably longer. It is flattened and the extremity has sharp edges above and below.

The limbs are stout. There are two distinct tubercles on the sole of the hind foot.

The ground color is a livid blue-black, brown, or black on the upper side. Scattered over the body, limbs and tail are numerous spots of bright yellow, which vary greatly in size, shape and arrangement. The spots may enlarge and become confluent, so as almost to cover the upper surface, or they may be almost obsolete. They are not limited to the upper part of the back, as they are in *A. punctatum*. The belly is of the same color as the upper surface, but it is usually wholly overlaid with sulphur yellow, so as to obscure the ground color. Through this wash of yellow may be seen the still brighter yellow of the spots. The chin and throat are often almost wholly yellow, yet this does not conceal the still brighter yellow spots. Prof. W. S. Blatchley has shown me a very large specimen from Terre Haute which is mostly yellow, but has on body and tail rather broad anastomosing dark bands. It suggests Cope's *A. xiphias*, but lacks the projecting lower jaw and the very long tail.

This species is the largest of the genus, individuals sometimes reaching a length of a foot. It is not uncommon to find them from six to nine inches long.

Distribution from Maine to Florida, California and Mexico. Indiana localities: Irvington and Indianapolis, rather common; Terre Haute

(Evermann, Blatchley), Crawfordsville (Beachler), Lafayette, very common (F. C. Test); in vicinity of Lake Maxincuckee; Wabash County (Ulrey). Probably to be found throughout the State. The specimens that I have seen from the northern portion of the State have the upper surface less spotted with yellow than usual.

This species has probably been more extensively studied than any other of the genus, and this especially on account of the fact that the gills and other larval structures are sometimes retained until the animal has reached the full adult size and even is capable of breeding. Dr. Elliott Coues (9, 4, 290), in speaking of specimens taken in North Dakota, says: "In all the specimens observed, the metamorphosis from the *Siredon* stage was completed at a length of four to five inches. In other regions I have procured the same species, still in the *Siredon* stage, but nearly twice as large." Some interesting observations have been made on the larvae of this species by Dr. R. W. Shufeldt (50, 1885, 263; 51, 453.)

Dr. P. R Hoy (22, v. 578) states that the Tiger Salamander, early in spring, about April 1, in Wisconsin, repairs to the water for the purpose of depositing its eggs. These are laid in packets of from twenty to fifty, on blades of grass. The eggs, including the gelatine, are a half inch in diameter, the yelk one-eighth. The tadpole emerges in twenty-five days, and has at that time a length of one-half inch. On each side of the mouth is a short "holder." By May 5 the "holders" are lost, the fore legs have made their appearance, and the larva feeds voraciously on aquatic insects. By the middle of August the gills have been absorbed. Thus about 100 days are occupied in attaining the adult condition. Dr. Hoy further says that when the feet and legs have been amputated, as they may be by water insects, they are reproduced, and the digits in the same order as originally.

Though I have frequently taken this species about Indianapolis, I have never been able to obtain its eggs nor to recognize the very young. In Indiana the metamorphosis is undergone when the tadpole is about four inches in length. I have seen many specimens taken at Irvington, and these will illustrate the peculiarities of the young at a time just preceding the metamorphosis. The entire length is 4.37 inches. There are present three external gills, each with numerous flat filaments arranged in two rows along the main axis. Three gill-slits are yet open, and these are guarded by gill-rakers similar to those of fishes. The teeth of the vomers and those of the palatines form separate patches, and in a series parallel with those of the maxillaries and premaxillaries. The tongue as yet shows no folds. The limbs are well developed. The tail has a broad, fin-like membrane above and below. The upper membrane extends forward well toward the head. The upper surface is extensively mottled and blotched with dusky. The color below is white.

On the back are three or four faint cross-bands of dusky. No indications are present of the future yellow spots so conspicuous in the adult.

In Indiana this species appears to pass the winter hiding about the margins of ponds, or, in some cases, away from the water, under logs and such places. I have received specimens taken under the ice, in company with *A. microstomum*, in January and February. Like the latter species, it mostly leaves the water as soon as oviposition is completed. They may then be found burrowing in the earth, when they can be found at all. However, this species, unlike *A. microstomum*, appears to be capable of remaining indefinitely in the water during the summer months and of enjoying its existence there. On the other hand, I have been told of a specimen that was taken in a dry corn-field on a hot day in August. A large specimen that was kept by me for several weeks seemed, during the warmer months, to prefer remaining covered up in a box of sand that was provided. At intervals it betook itself to the water. It was observed that this specimen shed its cuticle about every ten days. Before this exuviation occurred it entered the water and remained there for some time after the skin had been cast. During the colder months it preferred to remain constantly beneath the water, only coming up at intervals of fifteen minutes to take in air. Observations showed that this species, like *A. microstomum* and *A. punctatum*, enjoys an aquatic respiration. Water is steadily inhaled by the nostrils for five or six seconds, and then expelled by the mouth.

This species is a voracious eater, and will readily learn to take food from the hand. One was kept for weeks without manifesting any disposition to eat anything, but on being put into a cage along with an *Acris* and a large caterpillar, these mysteriously disappeared. One morning the salamander was caught holding a good-sized *Hyla versicolor* by one foot. Next morning the tree-toad was gone, while the salamander had an unusually bloated appearance. He would eat pieces of meat, angle-worms, and one day attempted to swallow a dead mouse.

Like all other animals of the kind, this one is regarded by many people as dangerously poison. It is, however, entirely harmless.

Ambystoma copeianum, Hay.

Cope's Salamander.

Amblystoma copeianum, Hay, 1885, 3, 209, pl. 14; Cope, 1889, 51, 63, with figures.

Head broad, body short, tail long and compressed.

The width of the head is a little less than the distance to the gular fold, and is contained in the distance from the snout to the groin 3.6 times. Distance to gular fold in distance to groin 3.2 times. Upper jaw projecting beyond the lower. Tongue like that of *A. tigrinum*.

Vomero-palatine teeth in four distinct series, made so by interruptions in the middle line and immediately inside the choanæ. The latter openings more widely separated than the outer. The body is short and depressed. The distance from the snout to the axilla is just equal to that from the axilla to the groin. There are eleven distinct costal furrows and a dorsal groove. The tail is equal to the distance from the snout to the beginning of the vent. It has a well developed crest and is compressed. Hind foot with two distinct tubercles.

The limbs are well developed, the posterior being a little longer, somewhat stouter, and the foot broader than the same parts in a specimen of *A. tigrinum* of the same size.

The color above is dark brown, almost black; below, brownish yellow. Between the fore and the hind legs the yellow color mounts up on the sides of the body to a level with the upper surfaces of the limbs. Head above like the back, below like the belly, with indications of a brighter yellow spot behind the symphysis.

This description is founded on the only known specimen, which was captured at Irvington, near Indianapolis. Nothing whatever is known concerning its habits.

Ambystoma punctatum, (Linn).

Spotted Salamander.

Lacerta punctata, Linnæus, 1766, *64*, i. 370; *Salamandra venenosa*, Holbrook, 1842, *54*, v. 67, pl. 22; *Ambystoma punctata*, Baird, 1850, *2*, 283; *Amblystoma punctatum*, Cope, 1867, *1*, 175; 1889, *51*, 56, with figures.

A species with broad head, stout body, black ground-color, and yellow spots.

Head depressed; widest at the swollen paratoid region. The greatest width is contained in the distance to the groin from 4 to 4.6 times. From the snout to the gular fold in distance from snout to groin 3.5 to 4 times. Gular fold not prominent, but rising high on the neck. Another fold from the angle of jaw running back to gular fold. Eyes of moderate size. Mouth large. Tongue moderate, with the papillary folds radiating from behind. Teeth of vomero-palatines in three portions, the extremities of the series being separated from the median portion by interruptions just behind the choanæ. The latter openings considerably further apart than the external nostrils.

Body plump, with a dorsal groove and 11 (occasionally 10) costal furrows. The distance from the snout to the axilla in distance from snout to groin 2.5 times. Limbs moderately developed; when placed along the side, just meeting or not quite. Toes rather short and depressed;

the plantar tubercles indistinct. Tail thick at the base, becoming compressed toward the tip, not high, and without a keel; a well-marked depression along each side; usually shorter than remainder of the animal. Skin smooth, but well furnished with mucous pores. A row of enlarged pores along the upper jaw, another inside the orbit, and another on each side of the upper edge of the tail.

The general color varies from slate-blue to deep black. There is an irregular row of spots along each side of the back and tail. Similar spots are found also on the head. From head to tail there may be from 10 to 20 of these spots. In alcoholic specimens these spots are white, but in life they are bright yellow. Those on the head are often bright orange, at least in spring. The spots are usually the size of the orbit. Under side of animal paler than above. May attain a length of seven or eight inches.

It is distributed from Halifax, N. S., to Wisconsin and south to Georgia and West Texas.

Indiana localities: Wabash County (Ulrey); New Harmony (Sampson's coll.); Wheatland (Ridgway); Franklin County (Hughes); Wayne County (Butler); Shelby County (G. H. Clark); Monroe County (Ind. Univ. coll.); Irvington; Terre Haute (Evermann and Blatchley). No doubt, exists throughout the State.

This species differs from *A. tigrinum* in having but eleven costal grooves, but a single series of yellow spots on the upper surface, in the fading of these spots to white in alcohol and in having no plantar tubercles, or but a single indistinct one.

Like its kindred, this species resorts in early spring to stagnant ponds for the purpose of depositing its spawn. They have been found about Irvington about the middle of March, hiding under pieces of fallen wood, which lie partly in the water. Later they disperse and may occasionally be found under logs. Dr. S. F. Clarke (*122*, 1880, No. 2) has studied the development of this species. The eggs are laid in masses of from 300 to 400. Each egg is covered with a thin coat of jelly, which swells up when brought in contact with the water. This is supposed to protect the eggs from fishes, but it probably protects them from many other enemies, animal and vegetable. This mass of jelly is much more solid than that of any other species of *Ambystoma* that I am acquainted with. While depositing her eggs, the female lies with her fore limbs extended laterally and her hind limbs curved around the opening of the cloaca, as if to assist in holding the eggs together. The male deposits the sperm on the eggs and thus fertilizes them. The egg has a light and a dark colored pole of equal size. Segmentation is most rapid in the light-colored pole. In due time the balancers appear, and are again lost on the thirtieth day. The larva has three pairs of gills, a tail with a fin-like membrane, and soon develops fore and hind limbs. Its transformation to the adult form occurs when a length of about two inches has been reached. (Cope, 51, 49.)

Like the related species, this salamander swims readily by vigorous strokes of its flat tail, while the limbs are held appressed to the sides. They are often found floating on the water of the aquarium. When disturbed they immediately plunge to the bottom and seek to hide. Their food-habits are closely like those of *A. tigrinum*. They devour with great greediness the angleworms that one offers them. The worms are swallowed by a succession of gulps. One was observed to swallow three inches of worm in five minutes. Prof. S. W. Garman has observed that the tail is prehensile and employed to prevent the animal from falling. I have observed the same thing.

This species, like *A. tigrinum* and *A. microstomum*, when under the water draws this in through the nostrils and at intervals expels it by the mouth. In this way they are enabled to remain for considerable periods under the water away from the air.

Ambystoma opacum, (Gravenhorst).

Marbled Salamander.

Salamandra opaca, Gravenhorst, 1807, *105*, 431; *Salamandra fasciata*, Holbrook, 1842, *54*, v. 71, pl. 23; *Ambystoma opaca*, Baird, 1849, *2*, 288; *Amblystoma opacum*, Cope, 1867, *1*, 173; *51*, 54 with figures; Boulenger, 1882, *28*, 40.

A species with a short, stout, swollen body, short tail, and weak limbs; the color dark, with light colored cross-bands.

The greatest width of the head is about three-fourths of the distance from the snout to the gular fold. The neck is distinct; the mouth extends half way to the gular fold. Tongue extensively free at the sides, and with the papillary folds radiating from behind. The vomero-palatine teeth consist of two lateral and a median series, the interruptions occurring just behind the choanæ.

The body has a swollen appearance. The length to the axilla is contained in the distance to the groin two and one-third times. There are eleven costal folds, and but slight indications of a dorsal groove. The limbs are but moderately developed. The animal has the appearance of being clumsy and weak. The tail is short and stout, its length equaling only two-thirds the distance from the snout to the groin. The plantar tubercle is wanting. The skin is everywhere pitted with minute pores.

The color is a dark brown or black. Across the back and upper side of the tail are a dozen light gray, or silvery white bands. These are broadest on the body. They usually fork on the sides and run together.

An irregular splotch on the head and nape of the same color. The limbs and belly may be uniform in color, but may be sprinkled with white dots.

Length of large specimens, 4.5 inches.

Distribution from Long Island and Florida, west to Wisconsin and Louisiana.

The following localities in Indiana have furnished specimens of this species: New Harmony (Sampson's coll.), Wheatland (Ridgway), Terre Haute (Blatchley). There seems to be no reason why a careful search should not be rewarded by finding this species anywhere in the State.

The habits of this salamander have been most carefully studied by Col. Nicholas Pike, with specimens taken on Long Island. He states (*48*, i, 209) that eggs and young were taken soon after the ice had left the ponds toward the latter part of March. The eggs were enveloped in a glairy mass similar to that of *A. punctatum*. The young emerged in fifteen days, but remained close about the glairy mass which they had escaped. Mr. Pike supposed that this furnished them food; this is, however, improbable. At first they are of a dingy brown color, with two rows of pale dots along the sides. When a month old, they were excessively active. Some which were dissected had in their stomachs the larvæ of insects, etc. At the age of two months, they would eat small mollusks. When an inch long the gills are fringed, the tail-fin is edged with black, the rows of white spots are more prominent, and the head broader and more prominent. The gills appear to be absorbed, and the fin membrane to disappear, when the length is about two inches. The whole body is described as being then sprinkled with white dots, as if flour had been thrown upon the animal. As soon as the branchiæ are absorbed, the larvæ become restless, seek to escape from the water, and if confined in it, many of them die. If permitted they crawl into moss and leaves, and curl up there in contentment. The metamorphosis occurs about the 5th of May. It is, however, not until the last of July that they assume the colors of the adult. From the time when the eggs are laid until the young have taken the complete adult form and color, there elapse about four and a half months, and the animal is then two and a half inches long.

Col. Pike regards the *A. opacum* as being strictly a terrestrial animal, entering the water only for the purpose of depositing its eggs. In confinement they refuse food for some time, and lie curled up, head to tail. At last they are willing to accept such mollusks as are offered them. It is entirely probable their food habits are similar to those of their relatives; and that they will eat almost any animal substance that they can swallow. Col. Pike states that this species hibernates late, hiding under leaves and burrowing in the ground. He says it has been known to burrow in soft ground to the depth of three feet.

Family V. PLETHODONTIDÆ.

(*Includes Prof. Cope's families Plethodontidæ, Desmognathidæ, and Thoriidæ,**
51, 33*).

Body salamandrine in form. No persistent gills or gill-slits. Vertebræ amphicœlous or opisthocœlous. Teeth on posterior edge of vomers. Parasphenoidal teeth present. Tongue extensively free on sides or all round. Carpus and tarsus cartilaginous.

KEY TO THE SUBFAMILIES OF *Plethodontidæ.**

A. Vertebræ amphicœlous. *Plethodontinæ*, p. 31.
AA. Vertebræ opisthocœlous.
    a. Carpus and tarsus cartilaginous. *Desmognathinæ*, p. 42.
    aa. Carpus and tarsus osseous. Extralimital. *Thoriinæ.*

Subfamily PLETHODONTINÆ.

Vertebræ amphicœlian. Carpus and tarsus cartilaginous.

KEY TO THE N. A. GENERA OF *Plethodontinæ.*

A. Tongue free along the sides, but not in front.
    a. Posterior digits 4.
        b. Costal grooves 18–31; Pacific States. *Batrachoseps.*
        bb. Costal grooves 13. *Hemidactylium*, p. 32.
    aa. Posterior digits 5.
        b. Mandibular teeth small, numerous, terete.
            c. Premaxillaries not ankylosed; costal grooves 10 to 19.
                                                                 Plethodon, p. 33.
            cc. Premaxillaries ankylosed; costal grooves 17; color, pale yellow. Georgia. *Stereochilus.*
        bb. Mandibular teeth, few, small, knife-shaped. Pacific States.
                                                                             Autodax.

---

*The three subfamilies of Plethodontidæ are founded on internal characters, and require some dissections. These, however, are not difficult to make. By making a short incision along the back of the specimen in hand, dressing away the muscular tissue down to the vertebral column, and then sharply bending the back so that two of the vertebra separate, it may be seen whether the anterior rounded head of the vertebræ is made of cartilage or bone. If it is of cartilage, the vertebræ are amphicœlous; if of bone, opisthocœlous. In either case, the posterior end of the vertebral centrum is concave. In like manner, the wrist and ankle may be dissected and the determination made whether the nodules found in them are composed wholly of cartilage or are bony. Since, however, we have no species of *Thoriinæ*, this examination is not necessary. It may facilitate the determination to recollect that the species of *Desmognathus* resemble in easily observed characters the species of *Plethodon* only, and the descriptions of these should be carefully scanned.

AA. Tongue extensively free all round and standing on a central stalk.
    a. Posterior digits 4. Southern States. *Manculus.*
    aa. Posterior digits 5.
        b. Premaxillaries not ankylosed. *Gyrinophilus.* Appendix.
        bb. Premaxillaries ankylosed. *Spelerpes,* p. 37.

Genus HEMIDACTYLIUM, Tschudi.

Hemidactylium, Tschudi, 1838, *99,* 54; Cope, 1889, *51,* 130.

No fontanelle between the parietals. Premaxillaries distinct. Tongue free along the sides, attached in the middle line in front. Limbs feeble; digits, 4-4.

Hemidactylium scutatum, (Schlegel).

Scaly Salamander; Four-toed Salamander.

Salamandra scutata, Schlegel, 1837, *106,* 119; *Hemidactylium scutatum,* Tschudi, 1838, *99,* 94; Cope, 1889, *51,* 130, with figures; *Batrachoseps scutatus,* Boulenger, 1882, *28,* 59.

Head flattened above, broadest just behind the eyes. Snout short, truncate. Width of head in the length to the groin about six times. Gular fold rising above nearly to middle line. Vomero-palatine teeth in two short series just behind the choanae. The parasphenoidal patches not in contact. Body cylindrical; the distance from snout to axilla in the distance from snout to groin three times. There is a dorsal furrow which runs forward to the head, and, there forking, sends a branch to each eye. Costal furrows 13 or 14. On each side of the back is a fainter longitudinal groove. Above this the costal furrows run forward and meet in the middle line at an acute angle. Base of the tail with a decided constriction, beyond which the tail again swells out and then tapers to a sharp point. Both upper and lower edges of the tail with an evident ridge for the greater part of the length. Skin of the whole upper surface granulated.

Limbs feebly developed; outstretched arms and legs about equal, and contained in the distance from the snout to the groin not quite twice. Fingers and toes 4, short, almost rudimentary.

Color above brown, chestnut, or purplish, mingled with pale spots and specks; snout, shoulders, limbs and upper surface of tail clay-colored. Below, the color is bluish-white, with many specks and small spots of black. The central line of the belly is spotted or not.

This species is distributed from Massachusetts and Canada westward at least as far as Illinois and south to Georgia. It is regarded as a rare animal, although it appears to be pretty abundant in places. Prof. Verrill has reported it to be common at New Haven, Conn. The Indiana localities from which it has been reported are: Brookville (Hughes), Terre Haute (Blatchley), Irvington (W. P. Hay), Wabash County at North Manchester (A. B. Ulrey). Mr. Hughes states that it was found in the moss at the roots of trees which stood near ponds. In this moss were also found the eggs. One of Mr. Hughes' pupils found about forty specimens, but did not secure them. The eggs are said to resemble those of that eminently terrestrial species *Plethodon cinereus*. Since the eggs were found near the water, it is possible that the larvæ spend a portion of their lives in that element. Prof. Cope says that the gills are absorbed at an early period of life, and he thinks that the animal never enter the water. The food is said to consist principally of worms and insects.

When this species has been dropped on its back, it will often lie for a time perfectly quiet, as if feigning death. I have heard it give a faint squeak like the scratching of a quill toothpick against paper. It can readily climb a perpendicular surface, and it can suspend itself by its tail. When thrown into water, it may hide for awhile at the bottom, but it soon endeavors to get out. The smallest specimen that I have seen is one from North Manchester, the total length of which is a little less than an inch and a half.

Genus PLETHODON, Tschudi.

Plethodon, Tschudi, 1838, *99*, 59; Boulenger, 1882, *28*, 53; Cope, 1889, *51*, 132.

Vomero-palatine teeth in two more or less oblique series which lie behind the choanæ. Parasphenoidal teeth present. Premaxillaries separated. Digits 4-5. Tongue free laterally, but attached medially in front.

Species belonging to North America seven. Of these, four occur on the Pacific Coast.

KEY TO THE EASTERN UNITED STATES SPECIES OF *Plethodon*.

Costal grooves 16 to 19. Color above ashy, or with a red dorsal band.
cinereus. p. 33.
Costal grooves 14. Color above black, with small white dots.
glutinosus. p. 36.
Costal grooves 13. Color above black, with large yellow spots.
æneus. Appendix.

Plethodon cinereus, (Green).

Ashy Salamander; Red-backed Salamander.

Salamandra cinerea, Green, 1818, *2*, 356; *Plethodon cinereus*, Tschudi, 1838, *99*, 92; Cope, 1889, *51*, 133; *Salamandra erythronota*, Green, 1818, *2*, 356; *Plethodon erythronotus*, Baird, 1849, *2*, 285; Boulenger, 1882, *28*, 57.

Form elongated and slender, with weak limbs. The length to the axilla in the length to the groin from 3 to 3.5 times.

The head is small and short, its width in the length to the groin seven times. The snout is short and rounded. Eyes large and prominent. Gular fold distinct, rising high on the sides of the neck. A distinct groove runs back from the corner of the eye to the gular fold, and is met by a groove rising from the corner of the mouth. The neck is distinct. Mouth large, the upper jaw slightly projecting. Tongue large, oval, free at the sides and slightly free behind. The vomero-palatine teeth in two short, arched, backwardly converging rows, which do not extend beyond the choanæ. The parasphenoidal teeth in two patches lying close together.

The body is cylindrical, without dorsal furrow, and with from 16 to 19 costal furrows, not including the one in the axilla, but including the inguinal. The limbs are short and weak, the outstretched hind legs scarcely equal to half the distance from the snout to the groin. The digits are short, the inner ones rudimentary. Tail equal to, or longer than head and body. Total length 3.5 to 4 inches.

Prof. Cope recognizes three varieties of this species, and these are based principally on differences of coloration. In all three forms the color below is whitish or yellowish, finely marbled with brown. On the sides the brown predominates, until it covers the surface, leaving only whitish specks. The middle of the back is variable.

Middle of back without a red stripe, ashy to black. *cinereus.*

Middle of back with a red or chestnut stripe, its borders parallel.
erythronotus.

Middle of back with dorsal stripe having dentate borders. *dorsalis.*

When the reddish or chestnut stripe is present it is somewhat broader than the inter-orbital space. Its central portion is usually finely marbled with brown. It may extend on the tail to its tip. In the form *dorsalis*, the stripe has deep indentations along its edges; these sometimes come opposite in pairs, when the band alternately expands and contracts. In other cases the indentations alternate, and then the band has a zigzag appearance. The variety *dorsalis* is furthermore said to have a shorter body than the others, the length to the axilla being contained in the length to the groin only 3 times, instead of 3.3 or 3.5 as in the other varieties. The number of costal folds is given by Prof. Cope as 16 instead of 18, as the others have. The variety *dorsalis* has been described from specimens collected at Louisville. I have two good specimens taken at Wyandotte Cave, not far from Louisville, which have the broad zigzag dorsal band and seventeen costal furrows, counting them as Prof. Cope counts them. In one the distance to axilla in distance to groin is 3; in the other 3.3. I think, therefore, that the form is hardly constant enough in its characters to be regarded as a variety. Mr. A. W. Butler reports this form from Bloomington.

The value of the *cinereus* and *erythronotus* forms is difficult to determine. They have been regarded as distinct species. Prof. Cope says that as varieties they are very permanent ones. He has found all of the young of the same brood or set of eggs, whether in the eggs or just escaped from them, uniformly with either dark backs or red ones. He has found red-backed specimens watching eggs with red-backed embryos, and brown-backed adults in charge of brown-backed embryos. He states also that there is some differences in the geographical distribution of the two forms. In general, however, the two are found in the same region, and it is no unusual thing to find both kinds under the same log. Further accurate observations are needed in order to settle this question. Blatchley (*94*, '91, 25), reports finding them in equal numbers at Terre Haute. This species is found in the territory extending from Maine, Ontario and Wisconsin south to Mississippi and Georgia.

In Indiana, specimens have been reported from New Harmony (*cinereus*, Hay); Franklin County (Butler); Monroe County (Ind. Univ. col.); Terre Haute (Evermann and Blatchley); Brookville (*cinereus* and *erythronotus*, *51*, 135 and 137); Crawfordsville (Beachler); Wyandotte Cave (*dorsalis*, Hay); Lake Maxinkuckee (*cinereus* and *erythronotus*, Dr. Vernon Gould); North Manchester, Wabash County, both forms (A. B. Ulrey). Some of the specimens from the last named locality have the dorsal streak of a more brilliant red than any others that I have ever seen.

The habits of this species appear to be wholly terrestrial. The eggs are laid in May, in damp situations, under stones, logs and leaves. The young are at first provided with branchiæ, but these are soon absorbed, and are probably not of much use to them at any time. The young are solicitously watched by one or the other of the adults. Smith (*18*, 725), states that he found specimens in the neighborhood of Vassar College, on the 6th of April. He adds that when it is disturbed it runs swiftly away, unless it is accompanied by the young, in which case neither these nor the adult attempt to escape. He thinks the adults feed the young. The eggs, the same author says, are laid in packets of from 6 to 11, and sometimes as late as June, while in the White Mountains the period of oviposition may be delayed as late as August. The food consists of insects and small snails. In one specimen collected, near Wyandotte Cave, I found a small shell allied to *Helix*, while in the stomach of another were the bodies of numerous ants. The species is quite active, running away when molested and hiding under the leaves. It can climb glass by applying the feet and abdomen closely to the surface.

Plethodon glutinosus, (Green).

Slimy Salamander.

Figure 2, Pl. 1.

Salamandra glutinosa, Green, 1818, *2*, 357; Holbrook, 1842, *54*, v. 39, pl. 10; *Plethodon glutinosus,* Tschudi, 1818, *99*, 92; Boulenger, 1882, *28*, 56; Cope, 1889, *51*, 139, with figures.

The body of this species is rather heavy for the genus. It is cylindrical or somewhat depressed, and with a very shallow groove along the back, but not extending upon the tail. Skin smooth and shining, pitted with numerous minute pores, which secrete a white sticky fluid. The length of the snout to the axilla in the distance from snout to the groin 2.75 times. Costal grooves 14.

Head of moderate width, its width in the distance to the groin six times. Snout rounded or truncate, the upper jaw projecting beyond the lower. Gular fold not overlapping; met by a groove from the eye. The latter organs large and protruding. Tongue large, the posterior fourth and the sides free; the papillary folds radiating from behind. Vomero-palatine teeth in two short, separated, anteriorly convex arches, which laterally pass a little beyond the choanæ. Parasphenoidal bands in close contact throughout and anteriorly removed from the vomero-palatines. The length of the parasphenoidal bands equal in length to the distance between the pupils. Choanæ as widely separated as the external nostrils.

Tail equal to or a little longer than the remainder of the animal; cylindrical in section and tapering to a point. Limbs moderately developed, the tips of the outstretched hind legs being contained in the distance from head to groin 1.5 times. The digits are short and depressed; the inner on fore and hind legs are small, but distinct.

The color above is black or blue-black. Along the sides are numerous whitish blotches about the size of the eye, and these are sometimes more or less confluent. On the back and upper head the spots are usually smaller and less bright. Under surface of the head and neck paler; of the belly bluish, with minute dots of white, which are not always the mouths of the mucous pores. Sometimes the posterior half of the tail is of a reddish brown color. In some specimens, when living, there are numerous spots of a brassy hue on the belly and under surface of the head, the tail with rounded dots of yellowish. Rarely there are no white spots anywhere on the body.

The largest specimens of this species that I have seen are 7 inches long.

The species is distributed from Maine to Wisconsin, and south to Texas and South Carolina. Indiana localities are: Brookville (Cope, *51*, 143, specimens sent to Nat. Mus. by Dr. R. Haymond); Terre Haute (Evermann and Blatchley); Monroe County (C. H. Bollman); Crawfordsville (specimens shown me by Mr. Chas. Beachler).

This species resembles the variety *laterale* of *Ambystoma jeffersonianum*, and to a lesser extent *Ambystoma microstomum*. It may be distinguished from both by the possession of parasphenoidal teeth.

This species, like its relative, *P. cinereus*, appears to be wholly terrestrial. It probably never enters the water, even for the purpose of laying its eggs, although Smith (*18*, 726) states that in Georgia it enters the water in April for the purpose of breeding. If the young ever have gills they are lost at a very early age. The species spends its life hiding under logs and stones, whence at night it comes forth to seek its prey. This consists of insects, and probably any other small animals that may fall within its reach. It is quite active while moving over the ground, although it can not run so rapidly as *Spelerpes longicaudus* and *Desmognathus fusca*. It moves with a sort of leaping and wriggling motion. It is especially remarkable for the development of prehensile powers in its tail. It will wrap its tail around one's finger and hang there for an almost indefinite time. Although given to living on land it shows no aversion to entering the water.

Prof. Cope (*51*, 142) says that it is found more abundantly in mountaneous districts, and haunts rocky localities as well as forest mold and fallen logs. He thinks that it prefers a cool climate. It appears, however, to be abundant in Southern Illinois, and I found numerous and large specimens in the low lands of Eastern Mississippi.

Genus SPELERPES, Rafinesque.

Spelerpes, Rafinesque, 1832, *107*, i, 22.

Vomero-palatine teeth in two series, which either converge backward without reaching the parasphenoids, or run transversely to the anteriorly prolonged parasphenoidal patches. These patches either separated or joined along the middle line. Tongue small, supported on a central stalk, mushroom-like. Premaxillaries ankylosed, their spines enclosing a fontanelle. Limbs moderately well developed; digits 4–5.

This genus, as limited by Cope, contains ten species, three of which belong to Mexico, the remainder to the United States. *S multiplicatus*, being found only in the Southwestern United States, is not here described.

Key to the North American Species of *Spelerpes*.

A. With 21 costal furrows; color, dark. *multiplicatus*.
AA. With 15 costal furrows; body stout; color, red, with black spots; vomero-palatines meeting the prolonged parasphenoids.
ruber. Appendix.

AAA. With 13 to 14 costal grooves; body slender; vomero-palatines not meeting the parasphenoidal patches.
  a. Tail considerably longer than rest of animal.
    b. Yellow, with black spots; tail with black cross-bars.
longicaudus, p. 38.
    bb. Red, with black spots on body and on tail.
maculicaudus, p. 39.
    bbb. Yellow, with a broad dorsal band and another each side of back.   <i>guttolineatus</i>. Appendix.
  aa. Tail not much, if any, longer than rest of animal, back with a median row of dots, and on each side a dusky line or band. Ground-color yellowish, above and below.
bislineatus, p. 40.

Spelerpes longicaudus, (Green).

Long-tailed Triton.

Figure 3, Pl. 1.

Salamandra longicauda, Green, 1818, *2*, 351; Holbrook, 1842, *54*, v, 61, pl. 19; *Spelerpes longicauda*, Baird, 1849, *2*, i, 287; Boulenger, 1882, *28*, 64; *S. longicaudus*, Cope, 1889, *51*, 168, with figures.

Body elongated and slender, with well developed limbs. Tail considerably longer than the rest of the animal.

Head flat, tapering gently backward, more rapidly forward to the rounded and projecting snout. Width of head in length to the groin 6 to 6.5 times. Eyes moderately large and prominent. Gular fold prominent, on the sides of the neck projecting backwards. Tongue boletoid and capable of protrusion. Vomero-palatine teeth in two short, curved series, which approach but do not meet each other backward; neither do they reach to the parasphenoidal patches. Externally they do not reach forward to the hinder borders of the choanæ. The parasphenoidals in two distinct bands, approaching in front, but diverging posteriorly.

Body somewhat flattened, with a dorsal groove and thirteen costal furrows. Distance from the snout to the axilla in the distance to the groin 2.6 times. The tail is low and flattened, about 1.5 times the length of the remainder of the animal, sometimes nearly twice as long, running out to a sharp point.

Outstretched hind limbs contained in the length to the groin about 1.3 times; toes and fingers short.

The color is usually a bright lemon yellow, sometimes increasing in depth to an orange. Below, the tint is paler and without spots. Above and on the sides there are numerous jet black spots varying in size from specks to spots as large as the eye. The largest of these may form a row running back from each eye to the pelvis, forming an interrupted band along the side. On the tail the spots are usually confluent into vertical bars.

The length of adult specimens is usually about five inches. I have a specimen from Pennsylvania that is 6.5 inches long.

This species ranges from Maine and Wisconsin south to Florida and Louisiana. In Indiana it has been taken at Waveland, Montgomery County (A. W. Butler); in Monroe County (Ind. Univ. coll.); "Caves of Southern Indiana" (D. S. Jordan); Harrison County (Prof. Hallett, with specimens). It is possible that the reports of this species from Monroe County and "Caves of Southern Indiana" refer to the *Spelerpes maculicaudus*, which is found there.

This is one of the most beautiful and interesting of our batrachians. Its brilliant colors, its graceful form, and its innocent habits demand for it attention and kindly protection. It appears to prefer for its haunts rocky ground and the fissures of caves. Prof. Cope says that he has never seen it in the water. Others regard it as to a great extent aquatic.

Harlan (*39*, 96) states that it inhabits the swamps of New Jersey. It undoubtedly enters the water in order to deposit its spawn. I have been able to learn nothing about its food or about its breeding habits.

Spelerpes maculicaudus, (Cope).

Hoosier Salamander.

Figure 4, Pl. 1.

Gyrinophilus maculicaudus, Cope, 1890, *22*, xxiv, 966, with figures; *Spelerpes maculicaudus*, Hay, 1891, *22*, 1133.

A species resembling closely *S. longicaudus*, but differing in form, arrangement of vomero-palatine teeth, and color. Head broader and flatter than in *longicaudus*, contained in distance to groin 5 to 5.5 times. The distance from snout to axilla in distance to groin 3.5 times against 4 times in *longicaudus*. Tail long and compressed, containing head and body 1.5 times. Costal grooves 13 or 14, one more than in *longicaudus*. Arrangement of vomero-palatine teeth different from that in *longicaudus*. In this latter, the series runs forward toward the choana and then turns outward behind it, not reaching as far forward as its hinder border. In *maculicaudus*, the series runs forward to a point in advance of the hinder border of the choana, or even to its anterior border, and then turns abruptly outward and backward, so as to produce the form of a hook.

The ground-color varies from orange to vermillion red. That of *longicaudus* being usually lemon yellow, sometimes becoming reddish brown or orange. The head and body are irregularly spotted with black dots about the size of the pupil, or larger. The tail is similarly spotted, but the spots do not incline to form vertical bars, as in *longicaudus*. The lower surface is uniform.

In a specimen that I have from Brookville, there is on each dorso-lateral region a row of black spots, one on each costal space. This row begins over the arm and runs back on the tail. In the middle of the back there is an irregular row of spots. In a specimen from Bloomington, the spots are more irregular.

Total length of large specimens, 6 inches; head and body 2.5 inches.

This species was originally described from specimens sent to Prof. Cope from Brookville, Ind. It was first taken there by Mr. E. R. Quick. Mr. Edward Hughes has reported the occurrence of the species in Decatur County. I have also seen specimens from the neighborhood of Bloomington, Indiana, some of which were contained in the collection of the State Normal School at Terre Haute. During the summer of 1891 Mr. W. P. Hay found one of these salamanders in May's Cave, near Bloomington, and another in Kern's Cave, about a mile southwest of Bedford, in Lawrence County. Both these specimens were found clinging to the walls of the caves about four feet from the water, and the one in the last mentioned cave was about a quarter of a mile from the entrance. They made no effort to escape, and both were detected by the gleaming of their eyes in the candle light. During the present summer Mr. Hay took a specimen of the species in a small cave near Wyandotte Cave. Dr. Stejneger, of the National Museum, has shown me a specimen of the same species, which has been sent to him from Barry County, in southwestern Missouri. This extends greatly the range of the salamander.

These salamanders are said to be more aquatic than is *S. longicaudus*. I kept two of them for some time in an aquarium, and they seemed to spend a considerable portion of the time out of the water. They have the power of climbing up a perpendicular glass, and they will remain sticking in such places for a long time. When the glass, to which they are adhering, is turned horizontal, they can remain sticking to the under side. They are extremely active, and when pursued, escape with great rapidity.

Spelerpes bislineatus, (Green).

Two-lined Triton.

Salamandra bislineata, Green, 1818, *2*, i. 352; *S bilineata*, Holbrook, 1842, 54, v. 55, pl. 16; *Spelerpes bilineata*, Baird, 1849, *2*, 9, 287; *S. bilineatus*, Boulenger, 1882, *28*, 66; Cope, 1889, *51*, 163, with figures.

A small species, seldom exceeding 3.5 inches long, and with the tail varying from a little less to a little more than the remainder of the total length.

The head is rather narrow, being usually contained in the length to the groin about seven times, occasionally only five times. The snout is short and rounded, sometimes abruptly so. The eyes are prominent.

The tongue is free all round, as is the case with all of the genus, and can be thrust far out of the mouth. The vomero-palatine teeth consists of two short, curved series, which backwardly approach each other without meeting, and which do not reach the parasphenoidal patches. The latter are separated from each other by a narrow space.

The body is elongated and slightly depressed. The distance from the snout to the axilla in distance to the groin nearly three times. There are nearly always fourteen costal furrows, occasionally but thirteen, rarely fifteen. There is a well-defined dorsal furrow. The limbs are rather feebly developed, the outstretched hind limbs are contained in the distance from the snout to the groin about 1.5 times The tail may be a little shorter than the head and body, or a little longer. It is compressed, keeled above, rounded below, and ends in a sharp point.

The color above is clear yellow, or yellowish-brown, sometimes merely yellowish-gray; below pale yellow, without spots Along the middle of the back there is a row of small dark spots, and these may even become confluent into a narrow line. Beginning at the eye, there is a dark-brown line which runs backward on the body and tail. The lower border of this line may be continued down on the side, gradually fading out, however, so that the whole side may appear to be occupied by a broad dusky band. Just below the darkest upper edge of the band, in the intercostal spaces, there may be spots of the ground color.

Size, usually about 3.5 inches, although a length of 4 inches may be attained.

This species is very widely distributed, being known from Maine to Florida, and west to Wisconsin, and possibly Louisiana.

Indiana localities are: Brookville (Hughes and Butler); Monroe County (Ind. Univ. coll.); Waveland, Montgomery County (Butler); Marion County, abundant at places along Fall Creek, (W. P. Hay); Terre Haute (Blatchley); Vigo County (Nor. Sch. coll.). Some specimens from Vigo County have the ground-color, above, a gray merely tinged with yellowish. The snouts of these specimens were also unusually blunt. Similar specimens have been taken in Marion County. Mr. C. S. Beachler reports the species from Waldron, Shelby County.

Like a number of other species, this seems to delight in living in close proximity to the water without spending the whole of its time in that element It is found hiding away under sticks and stones and among dead leaves about shallow streams and rivulets issuing from springs. It is very active and very slippery, so that when it starts away it is extremely difficult to catch it. A number of adults and larvæ were taken June 1, by W. P. Hay, near Fall Creek, in Marion County. The adults were concealed among the dead leaves on the borders of a spring, while the larvæ were found in a little pool not far away. The adults were extremely active, and ran rapidly by a kind of combination of wriggling and jumping. The young, which had attained a length of one inch and a half, had all the markings of the adults, except that the color was less yellow. The limbs were developed as in the adults. There were three pairs of gills, which the little animals held erect in the water. The tail also had a broad membranous fin. When the young were disturbed, they would dart through the water with a velocity that was surprising. I was not able to determine what they eat. One of the adults, when captured, had in its mouth the white larva of some dipterous insect. The young show a decided disposition to leave the water long before the gills are lost. Both old and young have the ability to climb up perpendicular surfaces. Some young were kept in a fruit jar in some water, and it was found that they were dying. All the water was poured out, except a very little, after which they fared better. One was found to have crawled half-way up the side of the jar, and was resting there in apparent content.

I have not been able to ascertain the character of the eggs of this salamander, nor how they are disposed of. They must be laid in the early spring.

Subfamily DESMOGNATHINÆ.

Vertebræ opisthocœlian. Carpus and tarsus cartilaginous.

This subfamily contains only a single genus, *Desmognathus*, belonging to which there are three species, only one of which is likely to be found in Indiana.

Genus DESMOGNATHUS, Baird.

Desmognathus, Baird, 1849, *2*, 285; Boulenger, 1882, *28*, 77; Cope, 1889, *51*, 194.

Vomero-palatine teeth in a short curved series behind the choanæ; feebly developed or entirely wanting. Parasphenoidal patches present. Premaxillaries ankylosed. Tongue extensively free laterally and behind, little free in front. Digits 4–5. Vertebræ opisthocœlian.*

Externally this genus has the characters of *Plethodon*. It is, however, distinguished from the latter by its vertebræ being convex in front, and by the coalescence of the premaxillaries into one piece.

A. With 14 costal grooves.
    a. Abdomen of uniform pale color. N. Y. and Pa. *ochrophæa*.
    aa. Abdomen marbled with light and dark.
                                                                            fusca, p. 43.
AA. With 12 costal grooves.
    Color black above and below. Va. to Ga.   *nigra*.

---

*See foot-note page 439; p. 31 of author's edition.

Desmognathus fusca, (Rafinesque).

Brown Triton.

Figure 5, Pl. 1.

Triturus fuscus, Rafin., 1820, *108*, March; *Salamandra quadrimaculata*, Holbrook, 1842, *54*, 49, pl. 12; *Desmognathus fuscus*, Baird, 1849, *2*, 285; Cope, 1869, *1*, 115; Cope, 1889, *51*, 194, with figures; Boulenger, 1882, *28*, 77.

Body rather heavily built and somewhat depressed. The distance from the snout to the axilla in distance from snout to groin three times, or nearly so. There are usually 14, rarely 13 or 15, costal grooves. A dorsal furrow commencing at the nape and ending over the vent.

Head flat and the snout rounded. The gular fold prominent, rising on the sides of the neck and the turning forward towards the eyes. Another fold, starting on the under jaw, crosses the corner of the mouth and rises on the side of the head. This furrow is crossed by one which starts behind the eye and runs back to the gular fold. The eyes are very prominent. Tongue moderate in size, free at sides and behind, little free in front. Vomero-palatine teeth not strongly developed, sometimes entirely missing. When present they form only a short series on each side. Parasphenoidal teeth in two patches which diverge backwardly. Lower mandible somewhat undulate, especially in the males; toothed to near the angle of the mouth. Width of head in distance to the groin about five times. The limbs are feebly developed, the fore and the hind limbs, when stretched along the side, not meeting by about four interspaces. Fingers 4, toes 5, joined at the bases by a narrow membrane. Tail about as long as the remainder of the animal, tapering gradually to a sharp point; its section at the base circular becoming flattened further back on the upper edge into a narrow fin.

The color of this species is quite variable. The adults of full size are usually dark above, with the belly paler, and with mottlings of brown. Along with the brown of the upper surface are shades of gray and pink; the young, and sometimes the half-grown, with ochreous spots along each side of the back, and these bordered more or less with black. In some specimens the whole middle line of the back is yellow or somewhat orange, brightest along the outer border, and divided along the middle line by a row of black spots. Such specimens resemble *Plethodon cinereus erythronotus*. These bright colors are sometimes retained by the full-grown specimens.

A variety of this species, *auriculata*, which has not been recognized in Indiana, is distinguished by having a series of small red spots along the sides, and often a red spot from the eye to the corner of the mouth. Has been taken near Cincinnati, Ohio.

The habitat of this salamander includes the territory extending from Maine to Indiana, Georgia, Mississippi, and southwestern Arkansas.

In Indiana it has been taken, up to this time, only at Brookville (Haymond, Butler, Hughes), and in Monroe County (C. H. Bollman); Richmond (F. C. Test); Decatur County (W. P. Shannon).

Where found at all, this species appears to be one of the commonest. It is to a very great extent aquatic. Prof. Cope says that "it lives chiefly among the stones in the many shallow rivulets and springs of the hilly and mountainous regions of the country. It prefers the rapid and shallow streamlets. Here it may be found under every stone, or its delicate larvæ may be observed darting rapidly from place to place, seeking concealment among mud and leaves." My experience with them is that, while in confinement, they do not, at least during the summer months, remain wholly in the water, but prefer to lie hidden in their tunnels in moss, with the head sticking out, so as to observe what is going on. When they are lying thus, if an insect is presented to them on the forceps they spring swiftly forward and seize and swallow it. I have had them to take flies and small spiders. One attempted to swallow a nearly whole red-legged grasshopper. They are extremely active and vigorous, and as slippery as eels, and it is with the greatest difficulty that they can be retained in the hands. When one is put on the floor, it escapes rapidly by a sort of combination of leaping and running. When held on the hand they will leap from it in the endeavor to escape. The tail is to some extent prehensile, and may be employed to keep the animal from falling.

Prof. Baird originally observed that this salamander lays its eggs in a string, and this is wrapped several times around the body of one of the pair, which then retires to a spot of concealment while the eggs develop. This curious habit, about which more needs to be known, has been confirmed by Prof. Cope. How long the care for the eggs continues is not known. The larvæ retain the gills for a varying period, but usually until the animal is half-grown.

Family VI. SALAMANDRIDÆ.

Body salamandrine in form. No persistent gills or gill-slits. Vertebræ opisthocœlous. Vomers each with a long palatine process reaching backward over the parasphenoid, bearing along its inner edge a single series of teeth. No parasphenoidal teeth. Carpus and tarsus ossified. Our single species red or olive above, with a series of red spots on each side of the back.

As here defined, including Professor Cope's families *Salamandridæ* and *Pleurodelidæ*. It embraces from six to nine genera, of which we have only the following:

Genus DIEMYCTYLUS, Raf.

Diemyctylus, Rafinesque, 1820, *108*, 5; Cope, 1889, *51*, 202; *Molge*, Merrem, 1820, *96*, 1885; Boulenger, 1882, *28*, 6.

Vomero-palatine teeth on inner margin of long backwardly projecting palatine processes of vomers. No parasphenoidal teeth. Tongue small, free only a little at the sides. An arch of bone connecting the squamosal and frontal bones. Digits 4-5.

Diemyctylus viridescens, Rafinesque.

Green Triton; Newt.

Figure 6, Pl. 1.

Triturus (*D.*) *viridescens*, Rafinesque, 1820, *108*, No. 22; *Salamandra symmetrica*, Holbrook, 1842, *54*, v, 57, pl. 17; *Diemyctylus viridescens*, Cope, 1889, *51*, 207, with figures; *Molge viridescens*, Boulenger, 1882, *28*, 21; *D. viridescens*, Gage, 1891, *22*, 1103, with plate and complete synonomy.

Body usually rounded and full, without a dorsal crest, but with a distinct sharp vertebral ridge. Distance from the snout to the axilla in distance to groin 2.5 times. Head longer than broad, its width in distance to the groin about 3.5 times. Outlines of head converging in front of the eyes to the rounded snout. Skull flat above, with four ridges enclosing three grooves; the outer ridge formed by the fronto-squamosal arch. Sides of the head perpendicular. A row of three or four enlarged pores behind the eye; these sometimes wanting. No gular fold. Tongue small and fleshy, free a little at the sides. Vomero-palatine teeth in two rows, which meet between the choanæ; then, diverging gradually, run backward along the roof of the mouth to the back of the skull. Tail constituting about one-half the entire length of the animal, more or less compressed, and tapering to a point; in the breeding season furnished above and below with a membrane-like fin.

Anterior limbs slender, the inner finger rudimentary. Hinder limbs stout, the inner and the outer toes small. In the breeding season, black callosities appear, in the case of the mature males, on the inner side of the hinder limbs, on the bottoms of the feet, and at the tips of the toes, those above forming transverse ridges.

The coloration of this species is so different at different periods of its life that a specimen seen while immature and again when full-grown would with difficulty be recognized as belonging to the same species. The gilled larvæ are olive of various shades, with black specks above; below, the color is almost uniform whitish. The metamorphosed, but still immature specimens are, on the other hand, of an orange color, varying to vermillion. The belly is somewhat paler. There may or may not be small black specks on the upper surface; while there are on each side of the back about half a dozen crimson spots of the size of the pupil, each surrounded by a black ring. Such young specimens are terrestrial in their habits. The sexually mature individuals are generally of some shade of olive, often having a reddish or yellowish tinge. The lower surface is generally of a bright yellow. Both above and below are to be seen numerous dots of black. In the breeding males some of these spots, especially those on the tail, may enlarge and become more or less ocellated. On each side of the back are the red spots surrounded by a black ring. The red, immature specimens are generally quite rough above and often below through the development, all over the skin, of small, pointed papillae.

The distribution of this species is from Maine to Hudson's Bay, Wisconsin, Texas, and Georgia. Indiana localities are as follows: New Harmony (Sampson's Coll.), Brookville, where it is common (Hughes), Mt. Carmel (L. M. Turner), Monroe County (Ind. Univ. Coll.), Terre Haute (Evermann, Blatchley), Rochester, Fulton County (Dr. Vernon Gould), Lake Maxinkuckee, Marshall County (Hay.)

The habits of this beautiful and interesting creature have been studied by many naturalists, among them Dr. Hallowell, Prof. S. F. Baird, Prof. E. D. Cope, Dr. Howard Kelly, Sarah P. Monks, Col. Nicholas Pike, and Prof. S. P. Gage. The red form is so different from the sexually mature, "viridescent" form that it was originally described as a distinct species under the name *miniatus*, and indeed was put by Rafinesque in a different subgenus. Prof. Baird placed the forms in the same genus as distinct species, but he recognized the close similarity existing between the two. Prof. Cope as early as 1859 (*1*, '59, 122-128), expressed the opinion that *miniatus* is only a state of *viridescens*; yet in his check list of 1875 he gives the two forms as subspecies. Later (*51*, 207) he regards them as "seasonal forms, which may be by reason of the environment rendered permanent for a longer or shorter time." Dr. Howard Kelly kept a number of the red *miniatus* in a dark box filled with wet moss and saw them transform into the olive state characteristic of *viridescens*. Sarah P. Monks observed the same change. Col. Nicholas Pike kept a number of the red ones where they could enter the water. After some time they did this, and after about three months they lost the bright red color, and in less than a year had the olive hue of *viridescens*. He, in common with many others, believed that the *viridescens* form might change back into the *miniatus* form.

Prof. S. P. Gage and wife of Cornell University have studied the species most carefully. They have seen the eggs deposited, watched the hatching of the larvæ, their transformation into the red, immature, *miniatus*, and the change of other specimens of these into the *viridescens*.

They believe that the terrestrial life continues until the autumn of the third or the spring of the fourth year after hatching, and that when they have once entered the water they do not again leave it unless the pool dries up or there is a scarcity of food.

The eggs are fertilized internally, and, when laid, are glued to the leaves of aquatic plants. Prof. Gage's observations seem to show that only about six or eight eggs are laid after each pairing of the sexes. After some days another batch may be laid. The time of deposition of the eggs extends from the first week of April until at least the middle of June. Gage's statements to the effect that the eggs are attached singly to the leaves of plants is confirmed by the observations of Cope and Monks. Prof. Verrill, of Yale College, and Col. Pike, on the other hand, state that the eggs are laid in masses containing from 25 to 150. The larvæ are active and timid. Their heads are not so broad as those of the larvæ of the Ambystomas, with which they may be associated. Their food consists of minute crustaceans, larval insects, and small snails. In color and form they closely resemble the full-grown individuals, except that they have three plumose gills, and the broad tail-fin extends forward to the back of the head. The gills begin to be absorbed during the last half of the August after hatching, the tail-fin is absorbed, and the larvæ then often comes to the surface to obtain air. About September 1 the now transformed young leave the water, hide away in the dirt and among leaves, and acquire the yellow or red color. During the metamorphosis the body appears to shrink considerably in size. I have seen, in the National Museum, a number of specimens, 2.25 and 2.75 inches long, with remains of gills, while red specimens, 1.75 inches long, entirely without any trace of gills are common. Prof. Gage, however, finds that the larvæ are usually only about an inch and a half long at the time of their change.

Where found in abundance, the adult newts may be taken during all the warmer months of the year; and it is probable that they are active during the whole year, unless the cold is too intense. They have been seen swimming under ice an inch thick. They delight most in pools which are fed by perennial springs. Their habits are not so nocturnal as those of many of their kindred. They may be seen at all times of day swimming about, climbing on aquatic plants, floating on the surface of the water, and basking in the sunshine. The food of the adults consists of insects, tadpoles, worms and mollusks. In confinement they become quite tame, and will take pieces of beef or insects from a wire, opening the mouth slowly, protruding the tongue, and gently pulling off the morsel. One has been known to swallow a piece of earthworm twice its own length, and to use its hands in holding its prey.

During the breeding season the colors are most intense, and the sexes are at their best. The spots along the sides become flame-color, the belly orange, while the male acquires a broad spotted crest along both edges of the tail. At the same time the black callosities appear on the hind legs. In their courtship the males are very active and very ostentatious. When success is won, they seize the female around the body and remain thus for an hour or more. Prof. Gage finds that the sexes pair in the fall as well as in the spring.

The outer skin is frequently shed. It is pushed back from the head by rubbing against objects; sometimes the hands are employed to effect this purpose. The process of moulting occupies about an hour and a half. Samuel Lookwood (22, x, 11) has seen it free itself of the cuticle while under the water; immediately the little thing turned around and swallowed the whole skin. Prof. Gage has seen the terrestrial form pull the exuvium off the end of the tail and swallow it. It is also interesting that the Newt can utter a faint shrill cry. The tail is extremely prehensile, and may be employed to suspend the animal for some time. It has been observed by Prof. Gage that during the aquatic stages the epithelium of the mouth is of the non-ciliated variety, while during the *miniatus* stage the epithelium is ciliated. Furthermore, the adults, while living in the water enjoy a form of aquatic respiration, water being regularly taken into the mouth and again expelled.

Order SALIENTIA.

Batrachia having a frog-like, or toad-like form. All four limbs present, the hinder greatly developed and fitted for leaping. Proximal elements of tarsus elongated, so as to form an additional limb segment. Vertebræ in front of the sacrum not exceeding nine. Ribs rarely present. No tail in the adult state. A tympanic cavity present.

This order includes the frogs and toads. With rare exceptions, the eggs are laid, fertilized, and hatched in the water. The young are familiar to all as tadpoles. These breathe for a short period by means of external gills, and for a longer time by internal gills. The intestines are long, the mouth furnished with horny sheaths, and the lips with several rows of horny teeth. By means of these, the little tadpoles are enabled to support themselves by scraping off the minute vegetable matter which covers objects in the water. As growth progresses, the limbs develop; the fore limbs, however, are for a long time concealed beneath the skin, and when they break through, they appear to be produced suddenly. As the period of metamorphosis approaches, the horny jaws and denticles are shed, the mouth enlarges, the tongue appears, the intestine shortens, the tail is absorbed, respiration by the lungs prevails, the gills disappear, and the tadpole leaves the water and becomes a frog.

All regions of the world, except the very coldest and the very driest, furnish representatives of this Order. Altogether there are about 800 species described. All our species come under Prof. Cope's suborders *Arcifera* and *Firmisternia*. Of the first, we have representatives of two families; of the second, one family.

KEY TO THE FAMILIES OF *Salientia*.

A. Clavicles* and coracoids of each side connected by an arched cartilage; that of the one side overlapping that of the other. (*Arcifera*.)
    a. Upper jaw without teeth; digits without disks.
                                                Bufonidæ, p. 49.
    aa. Upper jaw furnished with teeth.
        b. Form frog-like; toes and fingers with disks.
                                                Hylidæ, p. 52.
        bb. Form toad-like; digits without disks.   *Scaphiopodidæ*.
AA. Clavicles and coracoids of the one side firmly connected with those of the other side by means of a narrow median cartilage. (*Firmisternia*.)
    a. Upper jaw with teeth.   *Ranidæ*, p. 64.
    aa. No teeth on upper jaw.   *Engystomatidæ*.

Family VII. BUFONIDÆ.

Both upper and lower jaws destitute of teeth. Vomerine teeth usually absent. The diapophysis of the sacral vertebræ more or less expanded. Vertebræ procœlous. Ribs none.

A widely distributed family, containing, according to Cope, fourteen genera, of which we possess only one, *Bufo*.

Certain characters are very commonly possessed by the *Bufonidæ*. Among these are a heavy squat form, short limbs, a rough, warty skin, and a collection of integumentary glands lying behind the head, and known as the paratoids.

GENUS BUFO, Laurenti.

Laurenti, 1768, *109*, 25; Boulenger, 1882, *27*, 281; Cope, 1889, *51*, 260.

No vomerine teeth. Tympanum distinct or hidden. Toes webbed; fingers free. Sacral diapophysis more or less dilated. Outer metatarsals united.

---

*In order to understand the arrangements of the shoulder girdle on which the division of the Salientia into *Arcifera* and *Firmisternia* is based, the student ought to dissect this portion of the common Toad, as a representative of the *Arcifera*, and the same portion of the Leopard Frog, as an example of the *Firmisternia*. The accompanying figures, 7 and 8, pl. 2, will assist.

Bufo lentiginosus, Shaw.

Toad.

Bufo lentiginosus, Shaw, 1803, *71*, iii, 173; Boulenger, 1882, *27*, 308; Cope, 1889, *51*, 277 and 289, with figures. *B. musicus*, Holbrook, 1842, *54*, v. 7, pl. 1.

Var. *americanus*. *Bufo americanus*, LeConte, 1838, *53*, i, 75, pl. 9; 1842, *54*, v., 17, pl. 4; *B. lentiginosus*, var. *americanus*, Boulenger, 1882, *27*, 309; Cope, 1889, *51*, 284, with figures.

Var. *fowleri*, *B. lentiginosus fowleri*, Putnam, 1889, *51*, 279, with figures.

Form of body heavy and awkward. Head furnished above with more or less conspicuous bony crests. Of these, two, the fronto-parietal, run backward from the snout to the back of the head and embrace between them a furrow. Immediately behind the orbits each of these is met at right angles by a crest, the postorbital, which runs outward and downward behind the eye. From the lower end of the postorbital, a ridge, the supra-tympanic, may pass backward over the tympanic disk.

Snout short and blunt, the outline in front of the nostrils perpendicular. Tympanic disk distinct, oval, smaller than the eye. Upper jaw notched in the middle line. Parotoids elliptical, about as long as the distance from the nostril to the postorbital crest. Heel reaching to the eye or less. Toes half-webbed. Two metatarsal tubercles; the outer small, the inner large and with a cutting edge.

Skin on the upper surface everywhere provided with large and small pointed warts; below rough with crowded, pointed granulations.

Color variable; usually olive to brown, with irregular blotches and spots of dark brown; middle of the back with a light streak; below dirty yellow. The upper surface is sometimes almost a uniform black; at other times ash-gray, showing the dusky spots with great distinctness. Occasionally a specimen is found with the tubercles, and even considerable portions of the skin of a beautiful pink color. Other specimens may be of a uniform brick-color above, pink below.

Males provided with a large vocal sac, which communicates with the mouth by a slit on each side of the tongue. The toad may reach a length of as much as five inches, including head and body. Females larger than the males.

Four varieties, or subspecies, of this animal are recognized by Prof. Cope, three of which may be looked for within our borders. They may be distinguished as follows:

Fronto-parietal crests lying closely together, parallel, posteriorily passing little if any beyond the postorbitals. No supra-tympanic crest; a preorbital ridge. Heel reaching beyond the muzzle. Size rather small. Color grayish-olive; vertebral line yellow. *fowleri*.

Fronto-parietal crests diverging toward back of head, and passing somewhat beyond the postorbitals. The latter crests short; preorbital not strong; no supra-tympanic. Heel reaching the front of the orbit. Our common form. *americanus.*

Fronto-parietals diverging toward back of head, passing beyond the postorbitals and expanding into a pair of knobs; supra-tympanic well developed. Head from snout to ends of crests contained in length to vent 3.5 to 4 times. Southern. *lentiginosus.*

The common toad in its various forms has an extremely wide distribution, being known to occupy the whole of eastern North America, west to Montana, Arizona, and Mexico. In Indiana the variety *americanus* is everywhere distributed. The variety *lentiginosus* has not been reported from any locality within the State, but may be looked for in the extreme southern portion. The variety *fowleri* is known principally from specimens found at Danvers, Mass., but Prof. Cope reports a specimen from New Harmony. (See *51*, 279.)

The toad is with us an extremely common animal. It is almost a synonym for ugliness, but its mild and inoffensive disposition has gained for it some degree of toleration. Nevertheless, it still suffers much persecution, chiefly at the hands of untaught and thoughtless or cruel boys.

The toad appears in the spring when the warm days have fully come, and it is seen until the approach of the cool days of the autumn. Soon after emerging from its winter retreat, it repairs to the water for the purpose of depositing its eggs. These are laid, not as those of the common frog, in a shapeless mass, but in a long string, consisting of a double series of eggs enveloped in a tube of gelatinous materials. Mr. E. E. Crosby (*22*, vii, 574) says that the eggs of two specimens numbered respectively 8,840 and 2,200. Prof. Cope states that the young hatch early and are of a darker hue than is usual with other Salientia, but it is difficult to see how they can be blacker than the larvæ of *Rana pipiens*. The length of the young mature toads is about one-half inch; the color grayish, with small dark-colored spots. The metamorphosis occurs about July 10. The notes of the male toad are heard principally during the breeding season. They may be represented by the syllables ur-r-r-r-r.

The habits of the toad are mostly nocturnal, although it is not uncommon to see a toad hopping about in the daylight. Usually, however, they hide away during the day in holes and obscure corners, and come forth at evening to seek their food. This consists mainly of insects, and of these enormous numbers are devoured. It is related (*34*, '73, 23) that one old toad ate at one time twenty-three squash bugs, and on the top of these, ninety-four caterpillars. On account of this propensity for devouring insects, intelligent gardeners and farmers seek to induce toads to take up their residence on their grounds. No boy should be permitted to destroy this harmless animal. The prey is taken by suddenly projecting the tongue from the mouth, and then withdrawing it with the insect sticking to it. Besides insects, toads will eat earthworms, and small crustaceans, while it is said that one attempted to swallow a wounded humming bird. When the prey is large and difficult to swallow, the toad is said sometimes to push it against the ground or a stone. One is described as pressing its hand against its stomach, in order to hold down an earth worm that threatened to escape. The toad is stated never to make leaps after its prey, but to await until it has approached near enough to be reached by the tongue.

The warty skin of the toad is full of large glands, which secrete a thick whitish fluid. This has very acrid properties, and doubtless serves to render the animal unpalatable to most of its enemies. It does not seem, however, to protect it from snakes. It is said that this secretion will make the mouth of dogs sore, and even to cause some inflammation of the human skin. This last is doubtful. The skin of the frog is shed at intervals, and the statement has been made that this skin is immediately swallowed. During the winter the toad hibernates in holes and in the mud.

Dr. J. A. Allen, of Cambridge, Mass., tells of some toads that were taken in a torpid condition from the mud at the bottom of an old well. Some of them were buried in the mud two feet deep, and he supposed that they had been there ten or fifteen years, and probably longer. On being taken out, they soon revived and hopped away.

Family VIII. HYLIDÆ.

Upper jaw with teeth. Fingers and toes furnished with disks; these supported by the claw-like, terminal phalangeal bone. Transverse process of sacrum expanded. Vertebræ procœlous.

According to Prof. E. D. Cope, this family contains 183 species, and these he includes in 18 genera. They are distributed in all the great faunal regions. Of these genera, we have only three.

KEY TO THE GENERA OF *Hylidæ* FOUND IN INDIANA.

Fingers without a web. Toes fully webbed. Digital disks small.
Acris, p. 53.
Fingers webbed or not. Toes fully webbed. Digital disks larger.
Hyla, p. 55.
Fingers not webbed. Toes with little or no web. Disks small.
Chorophilus, p. 61.

Genus ACRIS, Dum. & Bib

Acris, Dumeril & Bibron, 1841, *74*, viii, 506; Boulenger, 1882, *27*, 336; Cope, 1889, *51*, 324.

Fingers free; toes fully webbed, the tips of the digits with small disks. Vomerine teeth present. Tongue with a notch behind. Tympanum indistinct. Sacral process little expanded.* Closely allied to *Hyla*.

This genus contains but a single species, and this is abundant everywhere with us.

Acris gryllus, (LeConte).

Cricket Frog.

Rana gryllus, LeConte, 1825, *62*, i, 282; *Acris gryllus*, Dum. & Bib., *74*, 507; Boulenger, 1882, *27*, 336; Cope, 1889, *51*, 324, with figures; *Hylodes gryllus*, Holbrook, 1842, *54*, iv. 131, pl. 33.

Variety *crepitans*.

Acris crepitans, Baird, 1855, *1*, 59; Boulenger, l. c.; Cope, l. c., 326, with figures.

Form frog-like. Length of head, measured to hinder edge of tympanic disk, in the length of body to vent three times. Snout pointed, its length in length of body six times; projecting beyond the lower lip. Vomerine teeth in two patches between the choanæ. Tongue broad, ovate, with or without a notch behind. Males with a large gular sac, which opens beneath the tongue. Tympanic disk seen with difficulty. Skin of the back smooth, or with small or large tubercles. Belly and thighs granulated; throat smooth. Legs long, the heel passing near to or beyond the snout. Two large metatarsal tubercles. Subarticular tubercles well developed. Fingers without web; toes webbed to near the tips.

Color variable and subject to rapid changes. Usually the upper surface is ashy-gray or brown. Occasionally green predominates, or there is considerable of reddish, especially along the middle of the back. Between the eyes there is a triangular spot of dusky, bordered with green in life. The middle line is usually pale. The upper lip is spotted with black. There is a blotch of dusky from the eye to the shoulder, and a stripe of the same above and behind the fore legs. Legs crossbarred Color below, white; under jaw often dusky. The length of large specimens is a little over one inch.

---

*FOOT-NOTE.—This character may easily be observed by making a transverse incision in the skin of the back at the point where the iliac bones are connected with the vertebral column; that is, where the frog's back bends abruptly. The transverse process cleaned of tissue, may then be compared with the same structure in the Common Tree-frog, *Hyla versicolor*.

Two varieties, or subspecies, of this frog are recognized, as follows:
Skin of back nearly smooth; hind foot from metatarsal tubercles longer than half the length of head and body. *gryllus.*
Skin of back considerably tuberculated Hind foot, from the metatarsel tubercles, shorter than one-half the length of head and body.
crepitans.

The habitat of this species extends from New York to Florida, and west to Nebraska and Texas. The variety *gryllus* is, for the most part, Southern in its range, while *crepitans* is more Northern. In Indiana the *crepitans* is found everywhere. The variety *gryllus* is in the National Museum from Mt. Carmel, on the Wabash, and I have specimens taken at Lake Maxinkuckee which correspond well to the descriptions of that variety.

This little frog is one of our commonest batrachians. During the summer season it may be seen in numbers along all of our streams. I doubt if it is often seen about the ponds or pools far from running water. It is not thoroughly aquatic, but delights to spend its time amid the vegetation about the border of the water. When alarmed it will leap into the water, but it often appears to become alarmed at its rashness and hastens to reach the land again. When followed up, however, it will go to the bottom and seek to conceal itself for awhile. Though belonging to the "tree-frogs" it never ascends trees, and probably climbs only the shorter grasses and water plants. It is a cheerful little creature, and on warm days may constantly be heard executing its noisy song. This resembles closely the striking together rapidly of two pebbles, and often, when their singing has been interrupted by the passer-by, it may be started again by clicking two stones sharply together.

This little chatterer appears very early in the spring, and it is my observation that it is to be found at all times during the summer. Numerous specimens were found at Irvington on the 8th of March. The eggs are probably laid about this time, although I do not know anything about them. On the 16th of August I found numerous specimens of the tadpoles of this species. They were found hiding in the vegetation at the bottom of a small stream. They were in very different stages of development, some with both the fore and the hind legs visible, others with only short hind legs. The arrangement of the horny denticles about the mouth of the larvæ I found to be different from that of specimens of *Chorophilus triseriatus.* In both species there are two rows of horny denticles on the upper lip. On the lower lip, there are in the *Chorophilus* three rows of denticles, but in *Acris* only two. Furthermore, in *Chorophilus* the denticles are finely serrated at their tips; in *Acris* this is not the case. The teeth of the latter genus are less numerous than in the former. The transformations occurred about September 1.

Dr. Holbrook states that the Cricket frog feeds on various kinds of insects, and makes immense leaps to secure its prey and to escape pursuers. He says that it can be easily domesticated, and takes its food readily from the hand. He kept several of them for months in a glass globe on a few sprigs of purslane, feeding them occasionally with flies; their chirp was at times incessant, and sprinkling them with water never failed to render them more lively and noisy.

Dr. C. C. Abbott (*22*, xvi, 707) has studied the habits of this species in New Jersey. He found them extremely abundant along ditches through meadows in the early spring, and very noisy. The eggs were deposited in little masses about May 1, and attached to blades of coarse grass. After the eggs were laid the number of frogs rapidly diminished, and by June 10 all were gone. Dr. Abbott believes that all die, and that those which appear later come from the transformation of the tadpoles. This occurs late in August. The frogs seem to be moderately abundant about the middle of September, but they are then found along rapidly flowing brooks. He also thinks that the frogs do not eat during the autumn, while in the early spring they are extremely voracious.

In this region I have found them during the whole summer, notably in August before the tadpoles of the year have transformed.

Genus HYLA, Laurenti.

Hyla, Laurenti, 1768, *109*, 32; Boulenger, 1882, *27*, 337; Cope, 1889, *51*, 349.

Fingers wholly, partially, or not at all webbed. Toes more or less webbed. The digits all provided with disks of larger or smaller size. Vomerine teeth present. Tongue entire or notched behind. Tympanum usually distinct, sometimes hidden. Sacral vertebræ, with its transverse processes more or less expanded.

The essential difference between this genus and *Acris* lies in the amount of expansion of the sacral diapophysis.

A large and widely distributed genus. North America possesses nine of the species. Of these seven occur in the Eastern United States.

KEY TO THE E. U. S. SPECIES OF *Hyla*.

A. Fingers one-third or one-fourth webbed.
    Gray, olive, or green above; a V-like mark between eyes and a large dark blotch on the back. Thighs yellow inside, and with dark mottling. Form stout.   *versicolor*, p. 56.
    Brownish or green above, without spots. A white line on upper lip and along the sides. Thighs not mottled or spotted behind. Form slender.   *carolinensis*, Appendix.

B. Fingers webbed not at all, or only the outer with a narrow web.
   a. Fingers entirely free.
      Olive-brown above, green in life; a stripe from eye along the side to femur. Toes two-thirds webbed. Tympanum one-third the eye. Rare. N. J. and S. C. *andersonii.*
      Olive or green above, with small, irregular dark spots. A V-shaped mark between the eyes. Toes one-half webbed. Tympanum one-half the eye. *squirella,* p. 58.
      Gray or olive above, with a X-shaped cross on the back. Thighs not spotted behind. Tympanum one-half the eye. Toes one-half webbed. *pickeringii,* p. 58.
   b. Outer fingers with a narrow web.
      Resembles *H. versicolor,* but with a web between the outer fingers and with the thighs behind dusky and with yellow spots. S. C. to Fla. *femoralis.*
      Purplish ash above, with numerous irregular dark spots. Limbs barred; thighs behind yellow, unspotted. Toes webbed to penultimate phalanx. Tympanum one-half to two-thirds the eye. Ga. to Fla. *gratiosa.*

Hyla versicolor, LeConte.

Chameleon Tree-Frog.

Hyla versicolor, LeConte, 1825, *62,* i, 281; Holbrook, 1842, *54,* iv, 115, pl. 28; Boulenger, 1882, *27,* 372; Cope, 1889, *51,* 373, with figures.

Form heavy and almost toad-like. Head considerably broader than long; the snout rounded; the space in front of the eye concave. Eyes large and protruding. Tympanic disk about two-thirds the diameter of the eye. Tongue large, circular, notched behind. Vomerine teeth in two closely approximated patches lying between, or a little behind, the choanæ. Chest crossed by a broad fold of skin. Fingers and toes with disks nearly as large as the tympanum. Fingers with an evident membrane uniting them. Toes webbed to near the tips. The first finger opposed to the rest. On extending the hind leg along the side the heel attains the back of the orbit.

Upper surface with numerous smooth warts; belly and under surface of thighs granulated. Males furnished with a large gular sac which opens on each side under the tongue.

The color in life varies from olive to dark gray and green, varying with the surroundings of the animal. Between the eyes there is a V pointing backwards. On the back there is a large, somewhat cruciform blotch of dark color, and behind it another smaller blotch on each side. The limbs are marked with broad bars of brown. The groin and the front and back of the thighs are bright yellow, over which is a network of brown lines. The gular sac of the males is often brown.

The length of a large specimen, head and body, is sometimes two inches.

This species occurs from Maine to Minnesota, and south to the Gulf and Texas. It is not necessary to specify localities in our own State.

Prof. Cope has recorded a variety of this species from Mt. Carmel, Ill., under the name of *phæocrypta*. "The color is a dark brown, with three rows of large approximated darker brown spots." Thighs yellowish brown, with little darker brown.

HABITS.—This species is by far our commonest representative of the *Hylidæ*. Although not often seen, it is heard by all, especially during the early spring. The time of appearance in the spring varies with the latitude and the weather. About Indianapolis I have heard them as early as the middle of April, and they may come from their winter quarters even earlier. Their breeding habits have been described by Mary H. Hinckley, as they are displayed in Massachusetts (*43*, xxi, 104 and 309; *22*, xvi, 636). The adults emerge from their places of concealment from May 1 to 10. The eggs begin to be deposited immediately, and this proceeds until July 4. The eggs are attached singly or in small groups to the grasses growing in the water near the shores. These eggs have extremely little gelatinous matter around them. The development of the tadpoles goes on with great rapidity, being accomplished within forty-eight hours. When hatched, the tadpoles are a quarter of an inch long, and of a pale yellow color, dotted on the head and sides with olive. Later the olive hue prevails, and there are markings of gold. The external gills appear and are resorbed during the first week. The "holders" disappear within the first ten days. The tadpoles do not huddle together. The hind legs appear at the age of three weeks. At that time the belly is iridescent, the back metallic, and the tail may be of a bright red. Soon, however, the color changes to a greenish, and by the time the metamorphosis has been completed, the color is a bright green. The arms appear during the eighth week. At this time they eat little food. They abandon the water from the 19th to the 24th of July, and their size is small. They betake themselves immediately to shrubbery and ascend it. They were seen to catch small spiders and to eat plant-lice. Some that were kept in a fernery gave evidence that they are nocturnal. Their green color is retained for about three months, when the frogs become gray. The black markings are not developed until after the young have forsaken the water.

The Tree-frog has a loud voice and may be heard during damp weather. It passes its time on trees, fences, and among vines, and it has the power of adapting its colors to the objects among which it lives.

The changes are said to occur slowly. These changes of color enable it to escape the notice of its enemies. These frogs appear to hibernate sometimes in the earth and sometimes in crevices in trees.

Hyla squirella, Bosc.

Squirrel Tree-frog.

Hyla squirella, Bosc, 1802, *57*, ii, 181; Holbrook, 1842, *54*, iv, 123, pl. 30; Boulenger, 1882, *27*, 398; Cope, 1889, *51*, 363, with figures.

A species of small size, the length seldom exceeding an inch and a quarter. The body and limbs slenderer than in either *H. versicolor* or *H. pickeringii*. The snout is somewhat pointed; eyes prominent; tympanum about one-half the diameter of the eye. Vomerine teeth between the choanæ. Head broader than long. Fingers without web; toes half-webbed. Heel reaching in front of the eye. Skin of the upper surface smooth; of the belly, thighs, and sometimes of the chin and throat, granulated.

Color above in life said to be green; below, white. In spirits, light brown or purplish. There is a V-shaped mark joining the eyes and several scattered spots of the size of the tympanum on the back. A dark band from the snout to the eye, thence through the ear to the shoulder. Below this there is a narrow light line. Limbs indistinctly barred.

The snout of this species is less projecting than in *H. pickeringii*; the tympanum larger and more distinct.

Until recently this species has been known only from South Carolina and west to Louisiana. Recently, however, a specimen was sent to Professor Cope by Mr. Amos W. Butler, from Brookville, Indiana. This gives a remarkable extension to the range of the species.

Of this frog Holbrook says that it is found on trees, often seeking shelter under the bark of such as are decayed. It frequently chooses old logs for the place of its hibernation. He had often seen them about old houses and under logs and boards.

Hyla pickeringii, (Storer).

Pickering's Tree-frog.

Hylodes pickeringii. Storer, 1839, *58*, 240; Holbrook, 1842, *54*, iv, 135, pl. 34; *Hyla pickeringii*, LeConte, 1854, *1*, 429, pl. 7; Boulenger, 1882, *27*, 399; Cope, 1889, *51*, 352, with figures.

Head a little longer than broad; snout rather pointed; the canthus rostralis distinct; the loral region concave. Interorbital space wider than the eyelid. Tympanum distinct, about half the diameter of the eye. Tongue free behind, rounded and slightly notched. Vomerine teeth in two patches, a little behind the choanæ. Males furnished with a large gular sac, which opens into the mouth by a slit each side of the tongue. Fold of skin on the breast conspicuous or not. Fingers and toes with well-developed discs. Fingers entirely without webs. Toes half webbed. Hinder limb, when pressed to side, bringing the heel to the eye. Two distinct tubercles on the heel; subarticular tubercles present. Skin nearly smooth above, granulated on the chest, belly and the under surface of the thigh.

Color above varying from ash to brown and reddish. On the back is an X of narrow dusky lines, these beginning at the eyes and terminating on the sides just before the insertion of the thighs. On each side of the body, opposite the intersection of the lines of the X, begins another dusky line, which runs parallel with the corresponding line of the X. Between the eyes is a very open V. A dusky band runs from the snout to the eye, and seems to be continued over the tympanum and the fore leg, and fades out along the side. Edge of the upper lip pale, but mottled with dusky. Some specks of brown on the breast and throat. Gular sac of male brown, at least in the spring. Fore legs of the color of the body, and with indistinct bars. Thighs mottled, tibiæ barred, feet mottled. The young frogs, with bodies three-fourths of an inch in length, are of a fine straw color, with the usual brown markings.

Length a little over one inch.

Known to occur from Maine (*18*, 707) to Manitoba and south to South Carolina and Indiana. Specimens of this frog have been taken in the vicinity of Bloomington by the late Charles H. Bollman. Professor Blatchley reports having taken four specimens at Terre Haute, and there was a specimen from there in the Normal School collection. I have it also at the hands of Mr. A. B. Ulrey, from North Manchester, Wabash County. I have also received a specimen from Brookville, at the hands of Mr. Ed. Hughes. It will, no doubt, be found ultimately in all portions of the State.

Prof. Cope states that this is our most abundant species of Hyla, but that it is more generally known by its voice than by its appearance· "After the rattling of the *Acris gryllus* in the marshes and river banks is fairly under way, during the first bright days of spring, the shrill cry or whistle of this little creature begins to enliven the colder swamps and meadows of the hill country. Different individuals answer each other with differently toned voices of a single note. This is exceedingly shrill and loud; the muscular force employed in expelling the air from the lungs seems to collapse the animal's sides till they nearly meet, while the gular sac is distended with each expulsion to half the size of the head and body together. They are chiefly noisy in the end of the afternoon, but in shady situations or on dark days may be heard through the morning and noon. When the breeding season is over they may still be found, but with difficulty, among fallen leaves in low places, where their color admirably adapts them for concealment, or in cellars, or on the ground in the woods. Not till the near approach of autumn do we have evidence of their ascent into the trees. Then, when the wind is casting the first frosted leaves to the ground, a whistle, weaker than the spring cry, is heard, repeated at intervals during the day, from one part of the forest to another, bearing considerable resemblance to the note of the purple finch, uttered as it is while flying. These voices are heard during the same season, that of the *Hyla* being distinguishable as slightly coarser, or more like a squeak. Both are associated with the weak chirp of the late *Dendrœca coronata* as it gleans its insect food on its southern flight. These are the latest sounds of autumn, and soon disappear before the steady advance of the ice king." (Cope, *51*, 354.)

DeKay (*30*, 69) says that this species is abundant in the neighborhood of New York, and that they are often to be seen on Indian corn and grape vines, and in greenhouses, under the leaves of plants during the heat of summer. He says that they feed on small flies; but it is not probable that they restrict themselves to such small creatures. Mr. S. P. Fowler sent the editors of the American Naturalist a beautiful fawn-colored specimen, which he had found imbedded in a heap of grass in his garden on November 25.

The eggs are laid in bunches of from 4 to 10, and are about one-twelfth of an inch in diameter. The eggs and larvæ in various stages of development are figured by Prof. Baird in Cope's "Batrachia of N. A." He states that the eggs are laid May 15. Smith says (*18*, 707,) that in Maine the eggs are deposited in April. According to Prof. Baird's statements, the young are able to swim within four days. For further details on breeding habits see Miss Hinckley's paper in Mem. Bost. Soc. Nat. Hist., vol. v, pp. 311–318.

Some larvæ of *pickeringii* which I have examined in the National Museum, at Washington, and which had the tail of full size, were a little less than an inch in length. Other young which had just transformed were only seven-sixteenths inch long. Prof. Blatchley reports finding this frog at Terre Haute, April 8, at the margin of a pond, to which they had evidently resorted in order to deposit their eggs. Two were kept in captivity several weeks, and regularly at 8 P. M. they began their piping notes, and kept them up about an hour. After this they were silent until the next evening. The gular sac would be inflated until it was two-thirds as large as the animal itself, when the air would be forced out, producing the notes. (*94*, '91, 27.)

Genus CHOROPHILUS, Baird.

Chorophilus, Baird, 1854, *1*, 59; Boulenger, 1882, *27*, 332; Cope, 1889, *51*, 331; *Helecetes*, Baird, *1*, 1854, 59.

Fingers free from web. Toes with little or no web. Digital disks all small, but the phalanx with a strong claw. Vomerine teeth present. Tongue round or oval, slightly notched behind. Tympanic disk distinct. Sacral vertebra with its transverse process slightly expanded.

This genus contains some six species, all except one belonging to North America.

KEY TO THE NORTH AMERICAN SPECIES OF *Chorophilus*

A. Upper jaw projecting beyond the rounded lower; profile pointed.
  Width of head in length of head and body 2.5 to 2.66 times; nostrils half way from tip of snout to eye. Texas. *ornatus*.
  Width of head in length 3 to 3.5 times; nostrils nearer tip of snout than orbit; no subarticular tubercles. Georgia to Texas.
  occidentalis.
  Width of head in length 2.8 to 3.66 times; subarticular tubercles well developed; upper surface usually with 3 to 5 dark stripes or rows of spots; nostrils nearer tip of snout than orbit. Eastern United States. *nigritus*, p. 61.
B. Upper jaw not projecting beyond the almost V-shaped lower; profile truncated; length two-thirds inch. South Carolina.
  ocularis.

Chorophilus nigritus, (LeConte).

Striped Tree-frog.

Variety, *nigritus*. *Rana nigrita*, Leconte, 1825, *62*, 282; *Cystignathus nigritus*, Holbrook, 1842, *54*, iv, 107, pl. 26; *Chorophilus nigritus*, Baird, 1854, *1*, 60; Boulenger, 1882, *27*, 333; Cope, 1889, *51*, 337, with figures.

Variety, *feriarum*. *Helecetes feriarum*, Baird, 1854, *1*, 59; *Chorophilus feriarum*, Cope, 1889, *51*, 339, with figures.

Variety, *triseriatus*. *Hyla triseriata*, Wied, 1839, *63*, 249; *Chorophilus triseriatus*, Cope, 1875, *12*, 30; 1889, *51*, 342; *C. septentrionalis*, Boulenger, 1882, *27*, 335, pl. 23, fig. 1.

Length of head and body seldom exceeding an inch and a quarter. Head having its length equal to its breadth or greater. Tympanum distinct, about one-half the diameter of the eye. Tongue emarginate behind. Vomerine teeth between, or slightly behind the choanæ. Males furnished with a conspicuous gular sac, which is capable of considerable inflation, and opening into mouth by a slit each side the base of the tongue. A conspicuous fold of skin across the breast. Fingers with, at most, a rudiment of a web. Toes with a slight web at their basal joints. Disks of all the digits small. Limbs of quite variable length, the heel reaching only to the tympanic disk, or to the nostril. Skin almost everywhere granulated, finely on the back, more coarsely on the lower surface.

General color, varying from light ash to fawn and purplish-brown; below, cream color. On the back and sides there may, in nearly all specimens, be seen five dark stripes or rows of spots. One beginning on the head, runs along the back, often forking on the urostyle. On the head, this stripe may expand so as to involve the eyelids, or there may be spots on the lids distinct from the median stripe. On each side of this dorsal stripe is another, which begins behind the eye and ceases just in front of the thigh. Another stripe begins on the snout, runs through the eye, over the tympanic disk and the shoulder, and fades out along the side. A whitish line runs along the upper lip to the shoulder. Sometimes the median and the lateral bands next it may be broken up into distinct spots, or these bands may be obsolete. The limbs are more or less barred or spotted with dusky.

Under the name *nigritus* I have included forms which are assigned by Prof. Cope to three distinct species, *nigritus, feriarum, triseriatus*. I do so because, at the present time, I do not think that the distinctions are sufficiently constant to characterize species. That the student may recognize the forms and accord them specific value, if he so desires, I state the differences.

Snout acuminate; width of head in length of head and body 2.8 to 3 times; heel reaching in front of orbit; size larger; color leaden or fawn, with three rows of dark spots above; these sometimes united into continuous bands. South Carolina to Mississippi. *nigritus*.

Snout shorter; width of head in the length 3 to 3.25 times; heel reaching to front of orbit; length of body in total length of hind leg, from 1.40 to 1.70; color ash or brownish, eyelids involved in median stripe, three parallel stripes above, seldom interrupted. Eastern United States to Illinois. *feriarum*.

Snout drawn out; width of head in length 3.5 to 3.6 times; heel reaching only to tympanic disk; length of body in total length of hind leg, 1.24 to 1.50 times; color ash to brown, with three parallel dark stripes, the median often forking behind; a distinct spot on each eyelid. New Jersey to New Mexico and Idaho. *triseriatus*.

C. nigritus is chiefly or wholly southern, *feriarum* eastern, and *triseriatus*, the short-legged and peaked-nosed form, mainly western, in distribution. All that I know of from Indiana belong to the variety *feriarum*. These have been taken at Brookville (Ed. Hughes), Brooklyn, Morgan County (W. P. Hay), Wheatland (Nat. Mus. coll.); Irvington; Terre

Haute (Blatchley), New Harmony and Mt. Vernon (Max. von Wied, *103*, xxii, 118); Wabash and Kosciusko counties (A. B. Ulrey).

This species appears to be rather rare in Indiana, but its rarity may be more apparent than real, since many of these animals have the faculty of effectually concealing themselves from the eye of the collector. In Indiana the variety *feriarum* lays its eggs about the 20th of March. The writer has described the life history of this little frog (*22*, xxiii, 770, with plate). Its eggs were found in a small pool on the 22d of March. They were attached to twigs in small and large bunches. Each egg was one-third inch in diameter, including the usual coating of jelly. In the egg the larva has a strong dorsal flexure, and has the tail thrown over the back. The tadpoles were set free on April 5. They are slenderer than the larvæ of the Leopard Frog, and not so dark in color. They are dark gray, rather than black. The external gills are small and quickly absorbed. They spend much of their time sticking to objects by means of their "holders," but very soon these organs disappear, and they then anchor themselves by means of their sucker-like mouths. The rudiments of the hinder limbs appear about the 20th of April. At the same time two rows of horny teeth appear on each lip, and a few days later an additional row on the lower lip. These teeth are minutely deticulated at their tips, and they form an admirable apparatus for scraping off the layer of nutritious slime that covers all objects in the water. There are from 55 to 95 of these teeth in each row.

As the tadpoles grow older the body becomes broader and the tail acquires a broad fin. By April 20 the length has become one half inch, and by May 4 about three-quarters. The body is of a dark color, adorned with numerous blotches of gold. The belly is nearly covered with a shimmer of gold and coppery. When three-fourths inch long the young were observed to come to the surface and to take in air.

By the 20th of May the young have attained the total length of a little over an inch. Many of them about this time succeed in releasing their forelegs from the skin which has held them down. Now the tadpoles grow smaller instead of larger. This is largely due to the shortening of the intestine at this period of transformation. These four-legged tadpoles are very lively and very timid. They show a great inclination to get out of the water and to hop about. They soon lose their skill in swimming, and if confined to the water too long will drown. The disks are seen on their fore feet as soon as these feet appear. The tails are rapidly absorbed, and by the 12th of June all have become little frogs like the adults, except in size. At the time of transformation the length of the head and body is less than one-half inch.

This species, though called a tree-frog, probably never climbs trees. They seem to live on the ground among the fallen forest leaves and in the grass. Prof. Cope states that it will leap into the water when alarmed, but will not stay there long. He also describes the note of *triseriatus*, which I have not heard. It is said to resemble that of the Cricket Frog, although not so loud. It may be imitated by drawing a point strongly across a coarse comb, holding them at the bottom of a jar and bringing them rapidly to the top. The same authority says that the note is uttered in the hottest part of the day.

Family IX. RANIDÆ.

Upper jaw furnished with teeth. Vomerine teeth present or absent. Transverse process of sacral vertebra little or not all expanded. Vertebræ procœolus. Ribs none. A large family, of some 20 genera and about 250 species. They belong chiefly to the Old World. Our species belong to the

Genus RANA, Linnæus.

Rana, Linnæus, 1758, *64*, ed. x, 354; Boulenger, 1882, *27*, 6; Cope, 1889, *51*, 393.

Teeth on upper jaw and on vomers. Tongue free behind and notched. Tympanum usually distinct, sometimes hidden. Fingers free. Toes webbed. Outer metatarsals separated by a web.

Contains more than a hundred species, living in all countries except southern parts of South America and New Zealand. Cope assigns thirteen species to North America.

KEY TO THE E. U. S. SPECIES OF *Rana*.

A. Without a black ear patch.

Dorso-lateral dermal folds present; heel reaching nearly to the muzzle or beyond it; back with well defined dark brown, pale edged oval or round spots. *pipiens*, p. 65.

Dorso-lateral dermal folds large, with smaller ones between; heel to front of orbit; tympanum one-half the diameter of eye; brown spots so large as to reduce ground color to a net-work of narrow lines; three phalanges of fourth toe without web.

areolata circulosa, p. 68.

Dorso-lateral dermal folds four; the *quadrate* spots of back in rows; two phalanges of fourth toe free of web; heel to front of orbit, or sometimes to muzzle. *palustris*, p. 67.

Dorso-lateral dermal folds present; tympanum one-half size of eye, or even larger than eye; hind foot longer than tibia or femur; large dark spots on back. *septentrionalis*. Appendix.

Dorso-dermal folds present; skin of back rough; tympanum nearly as large as eye, or larger; toes webbed nearly to tips; heel not reaching muzzle; dark blotches on back; size moderate.

clamata, p. 69.

No dorso-dermal folds; tympanum usually as large as eye, or larger; toes webbed to tips; usually some blotches above; size large.
catesbiana, p. 70.

B. Side of head with a black patch.
Head in distance from snout to vent 3.5 times; tympanum one-half the eye; skin of middle of back smooth; heel to middle of orbit.
cantabrigensis. Appendix.

Head in length 3 times; tympanum two-thirds the diameter of eye; skin of middle of back rough; heel to muzzle or more.
sylvatica, p. 71.

Rana pipiens, Gmel.

Leopard Frog.

Rana pipiens, Gmelin, 1788, *I.3*, 1052; *Rana virescens*, Cope, 1889, *51*, 397, with figures; *Rana halecina*, Schreber, 1782, *66*, 185; Holbrook, 1842, *54*, iv, 91, pl. 22; Boulenger, 1882, *27*, 41.

A very common, widely distributed and variable frog. Head varying in relative length, being contained in the length of head and body from 2.5 to 3.5 times. Vomerine teeth in two slightly oblique patches between the choanæ. Tympanum about as large as the eye. Heel reaching to the muzzle or a little beyond. A pair of prominent dorso-lateral folds starting behind the eyes and running to the end of the body. Between these may be two or more thinner folds. None of these folds are as broad as those of *R. palustris*. Another glandular fold runs from the corner of the mouth to the shoulder. No, or an inconspicuous, branch meeting this from the dorsal-lateral fold behind the tympanum.

Ground color ashy, olive or bright green above; below uniform white or yellowish. The upper surface with a number of rounded or oval brown spots of small to medium size, and these usually bordered with yellowish. The spots between the dorso-lateral folds are larger, and may be arranged in two or three rows, or may be irregularly placed. Outside these folds are two or three rows of smaller spots. The upper surfaces of the limbs are more or less conspicuously barred or spotted. Length of head and body of full grown specimens three to four inches.

Most of the males of this species have vocal sacs, which open by a small slit near the angle of the mouth. These sacs appear to be protruded through the slits, and serve to render the voice more powerful. Of this species Prof. Cope recognizes four subspecies, three of which may occur in our region. It must, however, be understood that forms intermediate are likely to be found.

Head long, contained in the length to vent 2.5 to 3 times; snout long and pointed; males with external vocal vesicles; heel reaching beyond tip of snout; web of foot leaving three phalanges of fourth toe free; spots of small size, scarcely bordered with pale; no longitudinal band on the front of femur. Mostly in Southern States. *sphenocephala*.

Head of medium length, contained in length 3 times; snout pointed; males with vocal vesicles; heel reaching tip of snout; web of foot leaving two phalanges of fourth toe free; spots of medium size, bordered with pale color; femur with a longitudinal band; tibial bars divided into two rows of spots. Maine to Mexico, but mostly in E. U. S. *pipiens*.

Head short, contained in the length 3.5 times; snout obtuse; vocal vesicles rudimentary; web of foot leaving nearly three phalanges free; spots larger and widely bordered with yellow; cross bars of tibia complete; no longitudinal bar or femur. Maine to Oregon and Mexico, but mostly Western U. S. *brachycephala*.

This species, as represented by its different varieties, is distributed all over the eastern portion of the United States and west to Oregon, Nevada and Mexico. The variety *sphenocephala* is chiefly southern in its range, *pipiens* (*virescens* of Cope) eastern and northern, and *brachycephala* western. Neither is confined, however, to these limits. With us the common form is *pipiens*, and it is everywhere abundant. *Sphenocephala* is in the National Museum from Wheatland, Ind., and to this I refer one specimen from Lake Maxinkuckee, which I find in the collection of the State Normal School. One specimen in the same collection and taken at Camden, Carroll County, had some of the characteristics of *brachycephala*. The head is contained nearly 3.5 times in the total length. Three specimens brought from Wabash County have the length of the head in length of head and body only 2.66 times. These might be referred to *sphenocephala*.

The Leopard Frog is our commonest, best known, and most beautiful frog. It is likewise the one that first makes its appearance when in the early spring the ice has relaxed its hold on the waters of our ponds and streams. Its cry is one of the earliest of the vernal notes. Indeed, it is no unusual thing, during an open winter, to hear its croak during the warmer days of midwinter. Its notes are of a varied kind. Prof. Cope says that its voice may be imitated by the syllables chock, chock, chock; but at times it has a cry that sounds like derisive laughter, and again, a sort of low, querrulous tone. Dr. Garnier (*22*, '83, 949) speaks of its "lugubrious and dismal love notes." Soon after it leaves its winter quarters, it proceeds to lay and fertilize the eggs. These are extremely numerous, and are laid in great masses, consisting of the small eggs enveloped in a large quantity of jelly. One pole of the egg is white, the other black. I have record of the deposition of the eggs at Irvington on March 18, and of the escape of the larvæ by April 1. The latter are five-sixteenths inch long; of this one-half is the broad tail. The numbers of these to be seen sometimes in ponds is astonishing. These tadpoles grow rapidly and transform during the first summer of their lives.

This frog is exceedingly active, and is capable of making leaps of 8 to 10 feet. They frequent ponds and streams, but they also leave these and travel to considerable distances in the search of suitable food. When thus traveling about they are much exposed to the attacks of snakes. Their intense green color must serve them well as a means of concealment from their enemies. Like other frogs they pass the winter buried in the mud at the bottoms of ponds and streams.

Rana palustris, LeConte.

Swamp Frog.

Rana palustris, LeConte, 1825, *62*, i, 282; Holbrook, 1842, *54*, iv, 95, pl. 23; Boulenger, 1882, *27*, 42; Cope, 1889, *51*, 406, with figures.

This species is very closely related to *Rana pipiens*, and especially resembles the variety *brachycephala* The head is relatively longer than in the variety last named, being contained in the length of head and body 3 times. The snout is usually obtuse, although not always so. Vomerine teeth in two nearly transverse patches between the choanæ. The tympanum is about two-thirds the diameter of the eye. The males have no external vocal vesicles, although Boulenger says that they have internal vesicles. The glandular folds of the back are especially well developed, being unusually broad. Of these the dorso-lateral are most conspicuous. They begin at the upper eyelid and run back to the hinder end of the body. Nearer the middle line and just behind the eyes two other folds begin, but appear to die out near the pelvic hump. Their places are there taken by two folds which lie close to the urostyle. All these folds are breader than in specimens of *Rana pipiens*. Above the tympanum the dorso-lateral fold gives off a branch which curves down behind the tympanum and meets another fold starting at the mouth and terminating at the shoulder.

The hind leg contains the length of the body 1.66 times. When the leg is pressed to the side the heel reaches the snout, or somewhat less. The web is deeply scalloped, and on the fourth toe leaves three phalanges free. The tubercles of the palm and of the sole are greatly developed.

The ground color is a pale brown or ashy above; below, yellowish white. Nearly the whole of the upper surface is covered with *squarish* spots of dark brown. The dorso-lateral folds are of the ground color. Within these there are two quite regular rows of the quadrate spots; the interspaces of these are much narrower than the diameter of the spots. Sometimes two or more of these spots coalesce so as to make a band. There is a spot on the snout and another on each eyelid. Outside of the dorso-lateral folds on each side is a row of spots like those of the back, but somewhat smaller. Lower down on the side is still another row of smaller and more irregular spots. On the upper jaw is a white streak from the snout to the shoulder. Upper and lower lips are mottled. The arms are spotted. The hind limbs are conspicuously barred from the groin to the toes. The posterior surface of the thighs is bright yellow, spotted with brown.

This species is found from New Brunswick west to Iowa, and south to Louisiana. It has been reported to me as occurring at Brookville (Hughes), and at Bloomington (Boliman). I have not seen the specimens from either of these places. In the State Normal School collection are two specimens that were taken at Lake Maxinkuckee. I have two specimens taken at Irvington, which I refer to this species with some doubt.

I have been able to learn little about the special habits of this frog. Prof. Cope tells us that it is a more solitary frog than *R pipiens*; that it prefers cold springs, and that it is seen in the grass more frequently than any others of our frogs. Its note is likewise said to be a low prolonged croak, somewhat resembling the sound of tearing some coarse material. Some authors tell us that it has a strong, disagreeable smell.

Rana areolata circulosa, R. and D.

Hoosier Frog.

Rana circulosa, Rice and Davis, 1878, *67*, ed. ii, 355; Cope, 1889, *51*, 413, with figures.

Rana areolata, as defined by Prof. Cope, is made up of four subspecies. Of these three belong to the Southern States, the fourth, *circulosa*, was originally described from specimens taken in Benton County, Ind. The species is characterized by a large, broad head, a short leg, imperfect palmation, and its peculiar coloration. The length of the head is one-third or more of the length, and its breadth is equal to, or greater than, its length. The leg, when pressed to the side, brings the heel only to the eye. The web is deeply scalloped and leaves three phalanges of the fourth toe free. Above, on each side, is a strong glandular fold, and between these there are several thin folds.

In *circulosa* the head is one-third of the length to the vent, and quite flat. The tympanic disk is oval or circular and only one-half the diameter of the eye. Between the dorso-lateral folds are six to eight narrow folds, and these may disappear in alcohol.

The striking thing about these frogs is their coloration. The spots found on the upper surfaces of their relatives, *R. pipiens* and *palustris*, have here become so expanded that they cover nearly the whole area. The ground color of yellowish is reduced to a network of narrow lines, or circles. Posteriorly in large specimens the lines of the net-work are broader and browner, and have a border of a lighter color. The spots themselves are of a reddish brown color. No light streaks about the head. Upper lip marbled with brown and yellow. Fore limbs mottled. Hind leg, including thigh, tibia, and foot, with wide bars of dusky and very narrow interspaces. Below, the color is uniform yellowish.

Found in northern portions of Indiana and Illinois. Described originally from a specimen which was found in Benton county, Indiana, by E. F. Shipman. About the size of *R. pipiens*, but one specimen in the National Museum is nearly five inches long from snout to vent.

Rana clamata, Daudin.

Green Frog.

Rana clamata, Daudin, 1803, *69*, 104; Boulenger, 1882, *27*, 36; Cope, 1889, *51*, 419, with figures; *Rana clamatans*, Holbrook, 1842, *54*, iv, 85, pl. 20.

Head broad and flattened, contained in the length of head and body about three times. Snout rounded. Eyes large and protruding. Vomerine teeth in two patches, the posterior edges of which project behind the line joining the hinder borders of the choanæ. Tympanic disk usually as large as the eye, sometimes larger; occasionally, especially in the females, somewhat smaller than the eye. A groove passes from the back of the eye over the tympanic disk and downward behind it, and terminates in front of the arm. There are present two dorso-lateral glandular folds. These start at the upper eyelids and run back to the pelvic region. Over the tympanic disk a branch is given off which passes down behind the disk and terminates over the arm. The groove described above lies between the disk and the branch. The skin of the back and sides is more or less rough. This roughness is produced by minute, sharp-pointed, wart-like elevations, those along the sides largest.

Hind limb, when pressed to the side, bringing the heel between the eye and snout. Third toe longer than the fifth. Toes webbed nearly to the tips. Subarticular and one metatarsal tubercles moderately developed.

Color above varying from greenish olive to brown; in life often bright green toward the head. On the back there are usually numerous small nearly circular blotches of dark brown, and larger ones on the sides. These are not spots, light bordered and definite, as those of *R. pipiens*, but look as if the color had run into the surrounding skin. The hind legs are crossed by narrow bars or rows of spots. The hinder surface of the thighs is granulated and of a yellow color, with spots and mottlings of black. The lower jaw and throat are marbled with brown, otherwise pale below. This species may be distinguished from *R. catesbiana* by the presence of the two dorsal folds. The length of head and body may be greater than three inches.

Distributed over the Eastern United States at least to the plains. Found all over Indiana in abundance.

This frog is thoroughly aquatic and lives along streams, in pools and in cold springs. It probably never leaves the water to travel any considerable distance. It is rather solitary, never collecting in such numbers as do the individuals of *R. pipiens*. They are timid, and when surprised by the stroller they leap, with a cry, into the water. Its voice may be represented by the syllable "chung." DeKay states that its note is more sonorous and in a lower key than that of the Bull-frog. The tadpoles of this species require two years for their development. Before the tail has been absorbed this species may be distinguished from larvæ of the Bull-frog by the presence of the dorsal glandular folds. When the fore limbs have appeared, the total length of the tadpole is about two and a half inches. The larvæ are stated to live on soft vegetable substances and never to be carnivorous.

Rana catesbiana, Shaw.

Bull-frog.

Rana catesbiana, Shaw, 1802, *71*, iii, 106, pl. 27; Boulenger, 1882, *27*, 36; Cope, 1889, *51*, 424, with figures; *Rana pipiens*, Harlan, 1825, *47*, 62; Holbrook, 1842, *54*, iv, 77, pl. 18 (not of Gmelin).

Attaining the largest size of any of our frogs, the body becoming in some cases 8 inches long, and the length of head, body and legs 18 inches. Head contained in the length 3 times or less; usually somewhat broader than long; sides of head sloping outward; snout rounded. Vomerine teeth projecting little behind the choanæ. Tympanic disk the size of the eye or even larger. Skin of the back smooth or minutely rough. Heel reaching to the front of the eye. Fourth finger slightly longer than the first. Third toe longer than the fifth. Toes webbed to the tips. There are no dorso-lateral glandular folds. A glandular fold begins at the upper eyelid, runs over and behind the tympanic disk, and in front of the arm, ending on the breast. Between it and the tympanic disk is a sharp groove.

The color of the upper surfaces in alcohol varies from reddish to olive and brown. In life the color may be pale yellow, green, brownish, or even deep brown. Below, the general color is white or yellowish. On the upper surface spots of brown may occasionally be almost entirely missing, but generally there are blotches of brown varying in size and number, sometimes running together. Their outlines are not well marked. The sides usually with distinct marblings of brown. The thighs may be spotted as well as the rest of the leg and foot. The lower surfaces may be almost devoid of any brown markings, or they may be conspicuously present on throat, breast, belly and legs. The hinder part of the thighs is usually mottled and blotched.

Distribution, Eastern North America to the Rocky Mountains. Everywhere in Indiana.

This species lives in the waters of our brooks, rivers and lakes. It probably never strays away from the vicinity of the water in the search of food. Its loud voice has given it its popular name of "Bull-frog." Prof. Cope says that its notes may be imitated by utturing a bass "br'wum" several times in a hoarse voice in front of an empty cask. Dr. J. H. Gardnier (22, 1883, 954) says that he has heard the Bull-frog at a distance of 5 or 8 miles. Authors tell us that when this frog is whipped it will cry much like a child; and Dr. Gardnier says that he has known the Bull-frog to hold its mouth open and scream for more than a minute, like a child in distress. These frogs are very voracious, and seem to catch and swallow almost any living thing that they can possibly devour. They have been known to eat cray-fishes, small fish, insects, worms, snails, mice, and even their own species. One was observed to be trying to swallow another of about half its size. Prof. J. A. Allen tells us that one seized and swallowed a cedar bird which he had shot, although the wings and tail continued to protrude out of the frog's mouth. Dr. Jos. Jones, of Georgia, reports having found in the stomach of one that he dissected a "grass-snake" three feet long. Other cases of the latter kind are on record.

The tadpoles of these frogs, like those of *R. clamata*, require two years for their complete development. It is possible that in some cases a greater number of years is passed in the larval state. A larva of this species was found to have a total length of 4.5 inches, of which 2.8 was tail. The general color above was dark olive green, below yellowish white.

This species is the one that is most relished as an article of diet. The frogs are often caught with hooks baited with red flannel.

Rana sylvatica, LeConte.

Wood Frog.

Rana sylvatica, LeConte, 1825, *62*, 282; Holbrook, 1842, *54*, iv, 99, pl. 24; Boulenger, 1882, *27*, 47; Cope, 1889, *51*, 447, with figures.

A rather slender and graceful frog, with a broad head and long legs. Length of head contained in the length of head and body from 3 to 3.5 times; its width greater than its length and contained in body and head

2.5 to 2.75 times. Snout rather pointed, the *canthus rostrales* distinct. Eyes prominent. Tympanic disk about two-thirds the diameter of the eye. A dorso-lateral glandular fold starts at the corner of the eye and continues along the side to near the vent. Over the tympanic disk it gives off a rather indistinct branch, which bends down behind the disk and terminates over the arm. There are no folds between the dorso-laterals. Another glandular fold begins at the corner of the mouth and stops just over the arm. The skin of the back, the sides, and upper surfaces of the legs is provided with numerous minute sharp points, which produce a slight roughness to the sight and touch. The hinder surface of the thighs somewhat granulated. Other surfaces smooth. Legs long, the heel reaching to the muzzle or beyond. Tibia longer than the femur. One metatarsal present. The subarticular tubercles feebly developed. Web leaving two phalanges of longest toe, and one of the others, free. Length of head and body may be about 3 or 4 inches.

In alcoholic specimens the color of the upper surfaces of body and limbs varies from pale reddish-brown to ashy or dark gray; the sides may be pale yellow or brown; beneath whitish. There may be a few indistinct spots on the pelvic region of the back, while the sides may be somewhat mottled with dusky. Lower surfaces occasionally indistinctly marbled. Limbs distinctly or indistinctly barred with brown. A dark stripe at the base of the humerus in front. A black stripe from the snout to the eye. A triangular brown ear patch. A white stripe from the snout and along the upper lip to the arm. Upper and lower lips marbled with brown and whitish. One specimen in my possession, captured at Irvington, has a light streak from snout to vent, reminding one of *R. calabrigensis*. In life the colors are variable. One was gray above, the back with a tinge of green, while the hind limbs had a tinge of reddish. The dark ear-patch was overlaid with a tint of copper. Iris golden. The glandular folds were golden-yellow, the flanks yellowish-green. Another living specimen was of a reddish or pink color everywhere above. In life the color is liable to undergo rapid and considerable changes, according to the surroundings.

This species occurs from Maine to the Athabasca River and south to South Carolina and Missouri. It is doubtless to be found everywhere in Indiana. Localities which have furnished specimens are as follows: Marion and Shelby counties (Hay); Franklin County (Hughes); Monroe County (Ind. Univ. coll.); Montgomery County (A. W. Butler); Hamilton County (F. C. Test); Terre Haute (Evermann and Blatchley); Wheatland (R. Ridgway); Wabash County (A. B. Ulrey).

This, it appears to me, is our most delicate, beautiful and interesting species of *Rana*. It is far less aquatic than the others, preferring to spend its life among the fallen leaves of the forest. It is not often seen, but occasionally occurs in enormous numbers. DeKay (*30*, 64,) says that Dr. Kirtland informed him that they were so abundant in the woods of Ohio that it was almost impossible to move without stepping on them. With us, at the present day, they are far less abundant. They are extremely active, and when pursued, they escape by making great and quickly repeated leaps. They are very skillful in hiding, and the close resemblance of their colors to the dead leaves and grass surrounding them renders it extremely difficult to find them. One that I was after leaped into the water and hid along side of a stone, and although I knew almost exactly where it was, it was some time before I could recognize it.

Miss Hinckley has studied the habits of the breeding adults and of the young (*43*, xxii, 91; *22*, xviii, 151). These frogs congregate very early in the spring for the purpose of oviposition, although they are not able to move in water below a temperature of 45° F. There may be as many as 1,380 eggs in one mass. The time of development varies greatly according to the temperature. The external gills were developed within two days after hatching, and in four more were wholly absorbed. The metamorphosis of the tadpoles was prolonged from April 12 to June 9. Tadpoles, which on May 31 measured 58 mm. in length, became reduced after the metamorphosis to 18 mm., about three-fourths of an inch. It was first observed by Prof. S. F. Baird that the tadpoles of this frog are carnivorous. He says that one way of preparing the skeleton of small animals is to put them in a vessel of water containing the living tadpoles of some of our frogs. These will devour the macerating flesh and leave the bones cleaned and hanging together by the ligaments. The larvæ of *R. sylvatica* are, he said, the most effective.

CLASS II. REPTILIA.

Members of the Class *Reptilia* pass their lives under extremely various conditions and circumstances. Many, as the larger number of the turtles, some snakes and the crocodiles, are given to haunting the waters of the sea, or of the rivers, lakes and ponds. Others, such as most snakes, and many tortoises, and the lizards, spend their lives on the land, and some of them in the hottest and most arid situations. Their eggs, even those of the most aquatic species, are always laid on the land; and the young, though they may, immediately after they are hatched, betake themselves to the water, never have gills, and depend wholly on their lungs for the oxygen that enters their blood. Their form is from the first like that of the adults.

In nearly all cases the epidermal layer of the skin of reptiles is disposed in the form of scales or plates. These may, on the one hand, be so small as to form mere granules, or, on the other hand, they may consist of a few large plates, like those of most tortoises. The soft-shell turtles, however, furnish us with an exception to the rule given. The epidermis of these is soft and moist, and is not broken up into definite areas. When limbs are present they are made up of the same elements as those of Batrachians, Birds and Mammals. But many reptiles, such as the snakes and many lizards, are entirely devoid of limbs. The blood of reptiles is cold, the heart has only three chambers (except in the crocodiles), and the arterial blood, as it is distributed to the body, becomes more or less mixed with the venous blood. As to their reproduction, most lay eggs; others bring forth their young alive, the eggs being retained in the body until the young have reached a considerable size.

KEY TO THE ORDERS OF EXISTING REPTILES.

A. Body covered with small, usually overlapping epidermal scales; these supported or not by bony plates of similar form and size, which also do not articulate with one another.
   a. Body usually extremely elongated; bones of upper jaws loosely connected with the rest of the skull; rami of lower jaw loosely connected at symphasis by elastic tissue. Serpents.
Ophidia, p. 75

b. Body shorter, but sometimes considerably elongated; limbs present or absent; bones of upper jaw firmly connected with remainder of the skull; mmi of lower jaw connected by suture. Lizards. *Lacertilia*, p. 131.

B. Body more or less protected by an armor of articulating bony scutes; these usually overlaid by firm epidermal plates of similar or different shape; absent, however, in some turtles.

    a. Body short and broad; encased in a bony shell formed above of the expanded ribs, below of dermal bones; four limbs; jaws covered by horny beaks; no teeth. Tortoises.
Chelonia, p. 142.

    b. Body lizard-like; more or less protected by transverse series of dermal bones; limbs four; jaws provided with teeth. Crocodiles and alligators. Extralimital. *Crocodilia*.

Order OPHIDIA.

Snakes.

Animals having a greatly elongated body and a tapering, pointed tail. Limbs wholly wanting, except in the rare cases where hinder limbs are believed to be represented by a pair of anal spurs. Vertebræ many and strongly articulated. Ribs movable and employed in locomotion. Braincase consisting of a reduced number of bones, which are firmly connected and form an efficient protection to the brain. Bones concerned in the seizing and swallowing of the prey—the palato-pterygoids, maxillæ, the quadrate and the squamosal—are loosely connected with the cranium and with one another. Mandibles loosely joined in front by elastic tissues. Gape of mouth usually large. No eyelids. Skin of upper surface provided with small, usually overlapping scales; belly with larger transverse scutes. Head usually covered with a few large, regularly arranged plates. Vent a transverse slit.

The *Ophidia* form a large and important order of Reptiles. They inhabit all the great faunal regions of the globe. They are few in species in the colder portions of the globe, but swarm in the hotter and moister regions. They are divided by naturalists into a large number of families, but of these we have representatives of only the following three:

KEY TO THE NATIVE FAMILIES OF OPHIDIA.

Maxillary bones elongated, with a row of several teeth in each; none of these grooved or perforated; no poison glands or fangs present.
Colubridæ, p. 76.

Maxillary bones each with a single erect, immovable poison fang, which is grooved in front. No pit between eye and nostril. Body with rings of black, red, and yellow. *Elapidæ*, p. 121.

Maxillary bones much shortened, bearing each a large, erectile poison-fang, with a few supplementary fangs; these perforated and the functional fang connected with the duct of a poison gland. A pit between the eye and nostril. *Crotalidæ*, p. 122.

Family I. COLUBRIDÆ.

Figure 9, Pl. 3.

Form elongated. Head more or less distinct from the neck. Tail tapering to a point. Head covered with large epidermal plates.* Maxillary long, furnished with a row of conical, solid teeth. No poison fangs or glands. Pupil circular. No pit in front of eye. No rattle. No anal spurs.

KEY TO THE GENERA OF THE FAMILY *Colubridæ*.

A. Anal plate divided.
   a. Dorsal scales not keeled.†
      b. Loral absent; antorbital present. *Tantilla.*
      bb. Loral present; antorbital absent.
         c. Nasal single, pierced by nostril.
            Rows of scales 13. *Carphophis*, p. 78.
            Rows of scales 19; two prefrontals.
                      Abastor. Appendix.
            Rows of scales 19; prefrontals united.
                      Farancia, p. 79.
         cc. Nasals 2, with nostril between them; rows of scales, 15 or 17. *Virginia*, p. 80.

---

*For plates illustrating the heads of North American species, see Volume X of the Pacific Railroad Survey Report.

†Sometimes in *Coluber* only the scales of the upper rows have keels, and occasionally all may be smooth.

bbb. Both loral and anteorbital present.
    d. Nasals 2; anteorbitals usually 2.
        Adult size large; pairs of subcaudal plates seldom fewer than half the number of ventral plates. *Bascanion*, p. 81.
        Adult size small; pairs of subcaudal plates seldom more than one-third the number of ventral plates. *Diadophis*, p. 84.
    dd. Nasal single; anteorbital single; subcaudals more than one-half the ventrals.
        Cyclophis, p. 85.
aa. Dorsal scales more or less keeled.*
    e. Rostral normal; not shovel-shaped or keeled.
      f. Three plates between the rostral and the eye.
        Nasal single; anteorbital and loral present. *Phyllophilophis*, p. 86.
        Nasals 2; anteorbital present; loral absent. *Storeria*, p. 88.
        Nasals 2; anteorbital wanting; loral present. *Haldea*. Appendix.
      ff. Four plates between rostral and eye (2 nasals, anteorbital, loral). Some of the outer rows of scales smooth; rows 25 to 29; ventral plates 200 to 270. *Coluber*, p. 90.
        All the scales keeled; rows 19 to 33; ventrals 125 to 160.
        Natrix, p. 95.
    ee. Rostral expanded and shovel-shaped, with a median keel. *Heterodon*, p. 102.

AA. Anal plate not divided.
    g. Dorsal scales smooth.
      h. Rostral plate normal. *Ophibolus*. p. 107.
      hh. Rostral enlarged, trihedral. *Cemophora*.
    gg. Dorsal scales keeled, rostral normal.
      i. Postfrontals two pairs; scales in 25 to 35 rows.
        Pituophis. Appendix.
      ii. Postfrontals one pair; scales in 17 to 21 rows; nasals divided. *Eutainia*. p. 112.
      iii. Postfrontals a single pair; scales in 19 rows; nasals single. *Tropidoclonium*. Appendix.

---
* See foot note on preceding page.

Genus CARPHOPHIS, Gervais.

Carphophis, Gervais, 1843, *72*, iii, 191; Garman, 1883, *13*, 99; *Celuta*, Baird and Girard, 1853, *6*, 120.

Small snakes with little heads and short tails. Head not distinct from the neck. Crown shields 7 or 9, there being in some cases but a single pair of frontals. Vertical broad. Loral present. No anteorbital. Nasals single. Scales smooth and glossy, arranged in 13 rows. Anal plates divided.

a. Color of the back descending below third row of scales.
amœna. p. 78.
aa. Color of the back not descending below third row of scales.
vermis. Appendix.

Carphophis amœna, (Say).

Ground-snake.

Coluber amœnus, Say, 1825, *2*, 237; *Celuta amœna*, Baird and Girard, 1853, *6*, 129; Cooper, 1860, *20*, xii, pt. ii, 302, pl. 19, fig. 2; *Carphophis amœna*, Gervais, 1843, *72*, iii, 191; Garman, 1883, *13*, 100, pl. vii, fig. 1; *Carphophiops amœnus*, Cope, 1892, *3*, xiv, 596.

Celuta helenæ, Kennicott, 1859, *1*, 100; *Carphophiops helenæ*, Cope, 1875, *12*, 34; *Carphophis helenæ*, Smith, 1882, *18*, 699.

Head small, the snout moderately elongated and rounded. Vertical hexagonal. Rostral convex. Postfrontals entering into the orbits, the prefrontals small or absent. Postorbital single. Upper labials 5; eye over third and fourth. Lower labials 6. Ventral plates 112 to 131; subcaudals 24 to 36 pairs. Scales smooth and glossy, arranged in 13 rows.

Color above, rich chestnut brown, below yellow to salmon. Length not exceeding one foot. The specimens without prefrontals have been regarded as belonging to a distinct species *helenæ*. Prof. Cope has shown recently, however (*3*, 1892, 596), that this is an inconstant character, and that the forms must be united.

The species is distributed from Massachusetts to Georgia, and west to Central Arkansas. Indiana localities are: Wheatland (Ridgway); New Harmony (Sampson's coll.); Monroe County (C. H. Bollman); Brown County (Chas. Jameson); Crawford County (Hay), where in the vicinity of Wyandotte Cave several specimens were found hiding under stones and logs; Terre Haute (Blatchley). Some of the specimens from New Harmony and those from Wheatland belonged to the form with prefrontals; all the others were without prefrontals.

Little is known about the habits of this innocent little serpent. DeKay states that it is found hiding under logs and stones. Mr. Sampson, of New Harmony, told me that he found it under the dead leaves in the forests. Its movements are probably nocturnal. Holbrook adds that it lives on insects. Dr. J. A. Allen states that it is decidedly subterrestrial in its habits, and is more frequently turned up by the plow or hoe than seen crawling on the surface. It seeks to bury itself when thus exposed. Observations on the breeding habits are needed.

Genus FARANCIA, Gray.

Farancia, Gray, 1842, *73*, 68; Baird and Girard, 1853, *6*, 123.

Head slightly distinct from the body. Crown-shields eight, the prefrontals being united. No anteorbital, the postfrontal and the loral entering into the orbit. Two postorbitals. Nasal single, grooved below the nostril. Scales not keeled; arranged in nineteen rows. Anal plate divided. Size large.

Farancia abacura, (Holbrook).

Horn-snake. Checkered-snake.

Coluber abacurus, Holbrook, 1836, *53*, i, 119, pl. 23; *Helicops abacurus*, Holbrook, 1842, *54*, iii, 111, pl. 26; *Farancia abacurus*, Baird and Girard, 1853, *6*, 123; *Hydrops abacurus*, Dum. and Bib., 1854, *74*, atlas, pl. 65; Garman, 1883, *13*, 36, pl. 1, fig. 5.

A snake reaching a large size. Head scarcely distinct from the body. Crown-shields normal, except that the prefrontals of the opposite sides are fused into one. No anteorbital. Loral and postfrontal forming anterior border of the orbit. Rostral low. Nasal single, groove below the nostril. Postorbitals two. Upper labials 7, the eye over 3d and 4th. Lower labials 8 or 9. Scales smooth and shining; arranged in 19 rows. Ventral plates 171 to 203; subcaudals 35 to 47. Anal plate divided.

The ground color may be regarded as blue-black. The sides are marked with about sixty transverse bands or wedges of bright red, which in some cases extend nearly to the middle of the back. These bands sometimes extend downward to the middle of the belly, and either join or alternate those of the opposite side. Since the red has definite margins and contrast strongly with the black, the belly has a checkered appearance. The head above is dark blue, with the plates tinged with red on their margins. Upper labials red, with a blue spot on each.

This snake reaches a large size, contrasting in this respect with its relatives, *Carphophis* and *Virginia*. One mentioned by Mr. S. Garman is 54 inches long, of which 5.6 is tail. This species is distributed from South Carolina to Louisiana and Central Arkansas, and up the Mississippi Valley to Knox County, Indiana. It and its eggs have been sent from

Wheatland to Mr. Robert Ridgway, of the National Museum. It has also been taken in Illinois just across the Wabash River from Indiana. I have seen it at Little Rock, Arkansas, in a cypress swamp. Holbrook says of this species that it is rare and shy, and that it lives in swampy ground and in damp places. It is a beautiful snake, as beauty goes among snakes.

Genus VIRGINIA, B. & G.

Virginia, Baird & Girard, 1853, *6*, 127; Garman, 1883, *13*, 96, in part.

Small, slender, and feeble snakes, with small head and short tail. Head narrow, rather high, snout pointed. Crown-shields 9. Postfrontals large, entering orbit and suppressing the anteorbitals. Loral present. Nasals 2, with the nostril in the anterior. Postorbitals 2, the lower small. Scales smooth, or feebly keeled on the posterior of body; arranged in 15 or 17 rows. Anal plate divided.

a. Scales in 17 rows. *elegans*, p. 80.
aa. Scales in 15 rows. *valeriæ*. Appendix.

Virginia elegans, Kenn.

Virginia's Snake.

Virginia elegans, Kennicott, 1859, *1*, 99; Garman, 1883, *13*, 98.

Originally described by Kennicott from specimens obtained "in heavily timbered regions in Southern Illinois." In my possession is a small specimen that was taken by the late Charles Jameson, of Indianapolis, at some point in Brown County. The total length is 5.75 inches. From this I draw a description.

Head small, narrow, and relatively high. Snout pointed, and the sides of the head perpendicular. Vertical hexagonal, with its right and left sides parallel. Occipitals large. Prefrontals entering the orbit and with the lorals forming its anterior border. Upper labials 6, the eye over the 4th and 5th. Scales in 17 rows; smooth, except that those on the tail are feebly keeled. Ventral plates 120; subcaudals 45.

Color gray, with a tinge of purplish, especially on the head. Below yellowish white, possibly reddish in life. On the dorsal surface are to be seen, here and there, small black dots. Those on the upper surface and the sides of the head more numerous and resembling points made by a fine pen.

From Brown county, Indiana, to Indian Territory (*10*, 84).

Nothing appears to be known about the distinctive habits of this delicate little creature. It must live on the smallest insects and worm-like creatures. Its subdued colors will undoubtedly enable it to escape the notice of its enemies. Its near relative, *Virginia valeriæ*, has not yet been found in Indiana.

Genus BASCANION, B. & G.

Bascanion, Baird and Girard, 1853, *6*, 93. *Bascanium,* Cope, 1875, *12*, 40.

Size large, form elongated; head distinct from the body; narrow and with sides perpendicular; tail long. Crown-shields 9. Vertical rather long and narrow. Two pairs of frontals. Loral present. Anteorbitals 2. Postorbitals 2. Nasals 2, with nostril between. Dorsal scales smooth; arranged in 15 or 17 rows. Ventral plates 170 to 210; subcaudals 80 to 150. Anal plate divided. Pairs of subcaudals seldom fewer than one-half the number of ventral plates.

Bascanion constrictor, (Linn.).

Black-snake. Black-racer. Blue-racer.

Coluber constrictor, Linnæus, 1758, *64*, ed. x, 216; Holbrook, 1842, *54*, iii, 55, pl. 11; Garman, 1883, *13*, 41, pl. iv, fig. 3. *Bascanion constrictor,* Baird and Girard, 1853, *6*, 93.

A long, slender snake, with a distinct head and a slender, whip-like tail, which constitutes about one-fourth the entire length. Head long, pointed, high; crown flat, and with the face bent down in front of the eyes. Eye in a groove which runs forward to the nostril. Snout pointed and rather projecting. Rostral high. Upper anteorbital large, the lower very small. Upper labials 7 or 8, the large eye over the third and fourth. Lower labials 8 to 10, sixth very large. Rows of dorsal scales in 17 (rarely 15 or 19) rows; the scales very smooth, the median narrow, the outer broad. Ventral plates 171 to 190; subcaudals 80 to 110.

The color of the adults is uniform above, but varies according to age and varieties from lustrous pitch-black to lead color and yellowish olive. Length 6 feet or more. In its varieties or sub-species this species has a range from the Atlantic to the Pacific and south to Mexico.

Variety *constrictor.*

Lustrous pitch-black above, varying to lead color. Below, the color is greenish-white or slate color, with the middle line paler. There is more or less white on the chin and the lower jaw. Upper labials with some white. Specimens from the prairies of the West and Southwest are of an olive-green color, and such shade into the variety *flaviventris.*

The colors of the young Black-snake are so different from those of the adult that one would hardly suspect it to be the same species. Instead of being of a uniform color above, they are much blotched and spotted. There is a series of reddish-brown blotches with black borders along the middle of the back, but disappearing on the tail. The blotches are about three scales long, and reach down to about the fourth row of scales

on each side. The sides are furnished with many specks and spots of
brown. The intervals between the spots are grayish or olive. The head
is mottled and specked. Below, the color is greenish-white, with three
or four specks of brown on each scale. Specimens over 18 inches begin
to assume the coloration of the adult.

The range of this variety is from the Mississippi River eastward. It
is to be found in all portions of Indiana, and it is unnecessary to specify
localities.

No species of our snakes is, probably, better known, or, at least, more
talked about, than is the "Black-snake" or "Black-racer," or, as it is
often called, the "Blue-racer." This is true probably because of its
abundance in all localities and because of its bold, active, and aggressive
disposition. It is, however, somewhat confounded with another common
Black-snake, *Coluber obsoletus*, which does not appear to be nearly so
active or so saucy.

The systematic name of this snake, *constrictor*, was given in allusion to
its supposed habit of entwining itself about the limbs and bodies of persons whom it might see fit to attack. Gmelin, one of its early describers,
speaks of it as running with great velocity, biting without poison, attack-
ing men by entwining itself about their limbs and squeezing. This is a
widely spread idea concerning these snakes at the present day, and there
is no doubt some truth in it. However, their daring is greatly exagger-
ated, and the stories that we may hear everywhere about their squeezing
people to death are without sufficient foundation. If one of these snakes
were driven to bay, or were seized, there is little doubt that it would defend
itself with great vigor and promptness. These serpents are not without
wisdom, and it is a common reputation which they have that they will
sometimes pursue persons who are retreating, but when the latter turn
the snakes will seek safety in flight. They are evidently full of curiosity,
and will often follow persons or objects, apparently merely to observe
them. I have been told of one of these snakes which was in a meadow
where a mowing machine was at work. The noise of the mower appeared
to excite the reptile greatly, and it followed the machine around the
meadow several times. At last it became so wrought up that it sprang
over the sickle bar and was cut into pieces. It is possible that in this
case the animal had young ones in the vicinity. Prof. Blatchley says
(*94*, '91, 31) that they are vicious, and will hiss and strike at a trespasser
when they are seeking a hiding place for the winter.

DeKay states that it is a bold, wild, and untameable animal, and that it
climbs trees with great ease by twining itself around the trunk in a spiral
manner. This it does in quest of eggs and the young of birds. Holbrook says that it feeds on mice, toads, and small birds. It is bold and
daring, entering barns and outhouses without fear, and has been known
to destroy young chickens. He also reports it as very irascible during

the breeding season, and that it will attack persons who may pass it, even at a distance of several steps. Its tail quivers with rage, making a quick, vibratory motion, which among dry leaves sounds not unlike the whir of the rattlesnake. It will even descend trees in order to attack an enemy who may tease it. He never knew one to try to twine itself about the legs, as it is commonly supposed to do.

Besides eating such creatures as have been already mentioned, the black-snake sometimes attacks and devours other snakes. Mr. F. W. Cragin (22, xii, 820) states that he found a black-snake swallowing a striped snake, *Eutainia sirtalis*, which he had killed the day before. The black-snake was 42, the striped-snake 22 inches long. This was a case in which the reptile was driven to partaking of cold victuals. Prof. A. E. Verrill, of Yale College, writes (22, iii, 158) that a student of Yale caught a large black-snake, and in bringing it home alive by the neck smothered it so that it became sick and vomited up a copperhead snake two feet long and in a nearly perfect condition. Soon afterward this was followed by a good-sized frog. Prof. Verrill supposes that the black-snake caught the copperhead while it was trying to swallow the frog; but this supposition is by no means necessary. Black-snakes are known to attack and destroy rattlesnakes in open fight. The black-snake is said to circle around the rattlesnake until the latter becomes confused or thrown off his guard, and then to spring suddenly upon the poisonous reptile, encircle him in its folds, and squeeze him to death. Dr. Elliott Coues (9, 4, 269) speaks of the hostility existing between the black-snake and the rattlesnake. In one case reported the black-snake threw two or three coils of its tail behind the rattlesnake's head and several others further back, and then, by a powerful muscular effort, tore the rattlesnake in two. When the black-snake has thus triumphed he has a right to a full meal. They are known to eat other species of snakes (22, ii, 136). When one snake swallows another the head is taken into the mouth first. A snake can swallow another almost as large as itself.

H. A. Brons, writing (22, xvi, 566) of the habits of some western snakes, says that this species and some others have the habit of swallowing whole eggs, and that it is no unusual occurrence to find such snakes with the entire contents of quails', prairie hens', and domestic fowls' nests in their capacious stomachs. With a little trouble they may be compelled to disgorge the ingesta unbroken. Miss Hopley, in her interesting book on snakes, says that these snakes will eat eggs, and that they will drink milk and eat cream. But when people tell us that they will suck the cows we must draw the line broadly and distinctly.

In the fall, and sometimes probably also in the spring, these snakes collect together often in large numbers, and we hear occasionally of "balls" of snakes having been seen. Brons, cited above, says that the female of the "Racer" is the larger, and not so graceful in form or movements. During the season of love making she seems to toy with the male, at times darting through the grass, among stones and into crevices to avoid him. On clear, level ground she is at a disadvantage. There, if she attempts to quit him, a coil of his tail is thrown about her body and his head laid upon her neck, and if it is removed, as promptly replaced, in the evident endeavor to propitiate her. Later in the season they are solitary or live in pairs.

The eggs of the black-snake are an inch and a half long and an inch in diameter. They are covered with a thick, tough skin. Out of such an egg I took a young snake 10.5 inches long. The snout was blunt, while a little sharp tooth projected from the middle of the upper jaw, beyond the lip. (See Agassiz *4*, 1, 288; Dr. Weinland, Proc. Essex Inst. ii, 28.) This is an "egg tooth," and its purpose is to enable the young to rip open the tough egg coverings when the time for hatching has come. This tooth is shortly afterward shed. Just the time of laying the eggs and the special way in which the female disposes of them I have not been able to learn. They are in all likelihood hidden away in soft earth or in rotten wood. In one female I found nineteen eggs, of which seven were in the left oviduct.

Genus DIADOPHIS, B. & G.

Diadophis, Baird & Girard, 1853, *6*, 112; Garman, 1883, *13*, 60.

Small, slender snakes, with a distinct and depressed head and a tail of moderate length. Crown-shields nine. Prefrontals two pairs. Loral present. Anteorbitals two. Postorbitals two. Nasals two, with the nostril between. Scales smooth, arranged in fifteen or seventeen rows. Anal plate divided. Ventrals 145 to 237. Subcaudals 36 to 60, seldom more than one-third the number of ventrals. Anal plate divided.

Diadophis punctatus, (Linn.).

Ring-necked Snake.

Coluber punctatus, Linnæus, 1766, *64*, ed. xii, i, 376; Holbrook, 1842, *54*, iii, 81, pl. 18; *Diadophis punctatus*, Baird & Girard, 1853, *6*, 112; Garman, 1883, *14*, 72.

A snake of small size, having a head distinct from the body, and a tail about one-fourth the total length. Head flat, snout rather broad and projecting beyond the lower jaw. Rostral low and broad. The lower anteorbital small. Upper labials seven or eight, eye over third and fourth, or fourth and fifth. Lower labials eight, fifth largest. Scales smooth, arranged in fifteen rows. Ventrals 148 to 203. Subcaudals 36 to 60.

The color above varies in the subspecies, or varieties, from olive through gray to blue-black; below from yellowish white to orange and red, with more or fewer dark spots. There is usually a light ring around the neck, close to the head.

The form usually found in Indiana is the typical *punctatus* The color above is a bluish black or a dark ash, with a wash of bronzy that extends down to the lowest rows of scales. Below, the color is orange or deep red, somewhat palest in front. On the outer ends of each of the ventrals there is a small black spot, and these are involved in the color of the dorsal scales. Near the middle line of the ventrals may be two rows of dark spots, or the spots on the ventrals may unite to form transverse bars. The ring around the neck is orange, edged with black. It is one or two scales in width. Upper labials yellow. The length may become about fifteen inches. This form is distributed from Nova Scotia to Georgia and the Mississippi Valley. Indiana localities are New Harmony (Sampson's coll.); Franklin county (Hughes); Monroe county (Bollman); Montgomery county (a specimen brought me by Mr. Beachler); Terre Haute (Blatchley); Shades of Death, Parke county.

HABITS.—Not much can be said concerning the habits of this little snake. Holbrook says that it is a very timid animal, living a great part of the time under the bark of trees, or old logs and stones. It emerges from its hiding places toward the dusk of evening, or after rain, when the insects on which it feeds have been washed from their hiding places. DeKay tells us (*30*, 39) that it is perfectly inoffensive, and that it emits a disagreeable odor. I can find nothing concerning its breeding habits.

Genus CYCLOPHIS, Günther.

Cyclophis, Günther, 1858, *26*, 119; Garman, 1883, *13*, 39; *Chlorosoma*, Wagler, 1830, *75*, 185; Baird and Girard, 1853, *6*, 108.

Form moderately elongated and slender. Head distinct from body. Tail long and tapering. Crown-shields 9. Loral present, small. One anteorbital, high. Postorbitals 2. Nasal single, nostril in its center. Eyes of moderate size. Mouth-cleft long and curved. Scales smooth; disposed in 15 rows. Anal plate divided.

Cyclophis vernalis, (DeKay).

Smooth Green-snake.

Coluber vernalis, DeKay, 1827, *2*, 361; Holbrook, 1842, *54*, iii, 79, pl. 17; *Chlorosoma vernalis*, Baird and Girard, 1853, *6*, 108; *Cyclophis vernalis*, Günther, 1858, *26*, 119; Garman, 1883, *13*, 39.

Body and tail rather long and slender, but less conspicuously so than in *Phyllophilophis æstivus*. Tail forming seldom more than one-third the total length, usually one-third or one-fourth. Head narrow and moderately high. Snout somewhat projecting beyond the lower jaw. Mouth-cleft large, curved. Upper labials 7; the eye over 3d and 4th. Lower labials 8; the 5th largest. Scales smooth; disposed in 15 rows. Ventrals 125 to 140; subcaudals 69 to 95; the latter seldom 70 per cent. of the former in number.

Color above, grass-green, fading somewhat on the lower rows of scales. Below greenish yellow. On the throat and upper labials yellowish white. The green of the upper surface often changing in alcohol to blue.

Distributed from Nova Scotia to Wyoming and southwest to New Mexico. Rare or not found in the Southern States.

In Indiana it is probably generally, but not abundantly, distributed. Known localities: New Harmony (Sampson's coll.); Brown county (collected by Chas. Jameson).

This is a beautiful and inoffensive little creature. Its color indicates plainly both that it lives among green plants and that it is little able to defend itself from the attacks of enemies. It must therefore depend on concealment for safety. It probably lives almost entirely in the grass, and rarely ascends trees and shrubbery, as does its relative, the Rough Green-snake. Holbrook says of it that "it is a very gentle animal, and can be handled with impunity; it seeks meadows of high grass, where crickets and grasshoppers abound, on which it feeds. It is found mostly on the ground, though at times I have seen it stretched on the branches of low shrubs, as the dwarf willow." DeKay (*30*, 40) says of it that it is exceedingly quick and lively in its movements; that it is most abundant in marshes, and that it is reputed to fight furiously with the Striped snake. Prof. F. W. Putnam found in Massachusetts, on August 31, its eggs under the bark of an old stump. They were just ready to hatch and one snake was already out. The eggs were an inch and a half long, and the young a little over 5 inches.

Genus PHYLLOPHILOPHIS, Garman.

Phyllophilophis, Garman, 1883, *13*, 40; *Leptophis*, Bell, 1826, *110*, 328; Baird and Girard, 1853, *6*, 106.

Body and tail very long and slender, and the body somewhat compressed. Head distinct from the body. Loral present, small. One anteorbital. Postorbitals two. Nasal single, with the nostril in the center. Eye large. Mouth-cleft deep. Scales keeled, except those of the outer one or two rows; arranged in 17 rows. Anal plate divided.

This genus differs from *Cyclophis* only in having the scales keeled.

Phyllophilophis æstivus, (Linn.).

Rough Green-snake.

Coluber æstivus, Linnæus, 1766, *64*, ed. xii, i, 387; *Leptophis æstivus*, Holbrook, 1842, *54*, iv, 17, pl. 3; Baird & Girard, 1853, *6*, 106; *Phyllophilophis æstivus*, Garman, 1883, *13*, 40.

Long and very slender; the tail whiplike, and constituting usually more than a third of the total length. Head separated from the body by a slender neck; narrow and high; swollen in the occipital region; the snout projecting considerably over the lower jaw. Eyes large. Scales keeled, except those of outer row and often some of the second row. Ventrals 150 to 165; subcaudals 110 to 135. The latter seldom as many as 70 per cent. of the ventrals.

Color grass-green above; below, greenish white. The green of the upper surface fades somewhat on the lower rows of scales. In alcohol the green changes to blue. The lower jaw and the upper labials are yellowish white.

Maryland to Kansas, south to Florida and Mexico. Indiana localities: New Harmony (Sampson's coll.); Vigo and Parke counties (Nor. School coll.); Dearborn county (A. W. Butler); Monroe county (Bollman); Cloverdale, Putnam county (Test).

HABITS.—This snake greatly resembles in general appearance and disposition the Smooth Green snake. It may be distinguished readily from the latter by its more slender form and by its keeled scales. It is equally as harmless as the other snake, and makes no attempt to bite when taken in the hands. It is given to climbing about on trees in search of the insects and larvæ that constitute its food. I have taken it while thus moving about on the branches of small trees. Of this species Holbrook says: "Perfectly harmless and gentle, easily domesticated, and takes readily its food from the hand. I have seen it carried in the pocket or twisted around the arm or neck as a plaything, without ever evincing any disposition to mischief. In its wild state it lives among the branches of trees and shrubs, shooting with great velocity from bough to bough, in pursuit of the insects which serve as its nourishment. Its green color, similar to the leaves among which it lives, afford it protection against those birds which prey upon it."

Prof. Cope (*22*, vi, 309) says of one that he kept in confinement that it manifested no disposition to climb over the ferns and plants among which it lived, but that it lived mostly underground. It had a habit of projecting its head and two or three inches of its body above the ground and holding itself for hours rigidly in a single attitude. In this position it resembled very closely a sprout or shoot of some green succulent plant, and might readily be mistaken for such by small animals.

Genus STORERIA, B. & G.

Storeria, Baird and Girard, 1853, *6*, 135; Garman, 1883, *13*, 29, in part.

Serpents of small size. Head distinct from the body. Tail only of moderate length. Crown-shields nine. No loral. One or two anteorbitals. Two or three postorbitals. Nasals two, with the nostril between. Scales keeled; arranged in fifteen or seventeen rows. Anal plate divided.

Scales in seventeen rows. *dekayi*, p. 88.
Scales in fifteen rows. *occipitomaculata*, p. 89.

Storeria dekayi, (Holb.).

DeKay's Snake.

Tropidonotus dekayi, Holbrook, 1842, *54*, iii, 53, pl. xiv; *Storeria dekayi*, Baird and Girard, 1853, *6*, 135; Garman, 1883, *13*, 31, pl. i. fig. 1.

Head somewhat larger than the neck, flat above, and rather high. Rostral as high as wide. Snout projecting beyond the lower jaw. One anteorbital, high. No loral. Two, sometimes three, postorbitals. Upper labials seven; eye over third and fourth. Lower labials seven; fourth and fifth large. Scales distinctly keeled; in seventeen rows. Ventrals 120 to 145; subcaudals 40 to 60. Tail one-fifth the total length.

The color of the upper surface is yellowish or reddish-ash, brownish-olive, or even chestnut. The middle of the back with a paler, clay-colored, dusky-edged band, three or four scales wide. On each side of this vertebral band is a row of brown or black dots about the length of two scales apart. These sometimes extend themselves and meet across the dorsal stripe. Occasionally the dots, and sometimes the band itself, are wanting. In such cases the color above is uniform. Below the dots mentioned, other dots are occasionally seen. The color of the lower surface is whitish or yellowish in alcoholic specimens, but in life the color is often salmon or red. The ventrals with one or two dots of brown at their outer ends. Plates of the head brownish, with some minute dots. In highly colored specimens, there is a large brown spot just behind the head on each side; another spot on side of head and across the corner of the mouth, and a small blotch under the eye. The length of grown examples is from twelve to fifteen inches.

Distribution from Maine to the Mississippi Valley, and south to the Gulf and to Mexico. Probably occurring in every township in Indiana. Known localities are: Wheatland (Ridgway); Lebanon (Nat. Mus.); New Harmony (Sampson's coll.); Harrison county (specimens from

Prof. Hallett); Monroe county (Bollman); Irvington; Terre Haute (Blatchley); Denver, Miami county, (Nor. Sch. coll.); Wabash county (Ulrey); Franklin county (Butler).

The colors of this little serpent are such as harmonize well with its usual surroundings, the soil and dead grass, leaves, and slender, broken branches of trees. I have observed no evidences of its being aquatic, but some observers make such statement. DeKay reports that all that were seen by him were either in the water or in the vicinity of it. One taken by him was swimming across a bay of Long Island Sound. All that Holbrook has to say about it is that it frequents meadows and places where the grass is of luxuriant growth, and feeds on various insects, as crickets, grasshoppers, etc. They are ovoviviparous. A female, taken at Cumberland Gap, Tenn., and in midsummer, contained eleven eggs. The eggs were .37 inch by .25.

Storeria occipitomaculata, (Storer).

Storer's Snake.

Tropidonotus occipitomaculatus, Storer, 1839, *76*, 230; *Storeria occipitomaculata*, Baird and Girard, 1853, *6*, 137; Garman, 1883, *13*, 30, pl. 1, fig. 2.

Averaging smaller in size than *S. dekayi*, which it resembles in proportions and in coloration. Snout short and blunt. Anteorbitals 2. Postorbitals 2. Nasals 2, with the nostril mostly in the anterior. No loral. Upper labials 5 to 6, growing larger posteriorly, the eye over third and fourth. Lower labials 6 or 7. Scales keeled; arranged in fifteen rows. Ventral scales 117 to 128; subcaudals 43 to 50.

Color olive to reddish gray or chestnut brown. Along the back there is a paler stripe about three scales wide, and this is usually edged with dusky. Bordering the pale vertebral band and situated in the dusky border in each side is a row of brown dots. The vertebral band and the dots may all be faint or entirely absent. Often there is a yellowish stripe on the lowest row of scales. Head like the body, but mottled with brown. Behind the occipital plate is a spot of yellow, salmon in life, and a similar spot on each side just behind the corner of the mouth. The fourth and fifth labials with a small spot of similar color. Below, the color is yellowish; in life salmon or brick-red. The ends of the ventrals mottled with dusky.

Length of grown specimens about one foot.

The territory occupied by this animal extends from Maine to Wisconsin and south to Georgia and Texas. In Indiana it is doubtless to be found everywhere. It has been collected at the following localities: New

Harmony (Sampson's coll.); Montgomery county (A. W. Butler); Lebanon, Boone county (S. F. Baird); Irvington, where it is less common than *S. dekayi*; Terre Haute (Blatchley).

I know little about the habits of this snake. Smith (*18*, 698) states that they are somewhat nocturnal, and live chiefly under logs and stones. They are in all probability ovoviviparous. In the stomach of a specimen taken at Irvington I found a slug.

Genus COLUBER, Linn.

Coluber, Linnæus, 1758, *64*, ed. x, 216; *Scotophis*, Baird and Girard, 1853, *6*, 73; *Elaphis* Garman, 1883, *13*, 53.

Snakes attaining a large size. Head distinct from the body; rather narrow and long. Crown-shields 9. Vertical broad. Loral present. Anteorbital 1, large. Postorbitals 2. Nasals 2, with the nostril between. Mouth deeply cleft, the outline nearly straight. Dorsal scales keeled, except some of the lower rows, which may be smooth; arranged in 23 to 29 rows. Anal plate divided. Ventral plates 200 to 240. Subcaudals 63 to 95.

(In rare cases, especially specimens of *C. guttatus*, all the scales may be smooth.)

ANALYSIS OF THE SPECIES OF *Coluber*.

A. With longitudinal bands of brown. N. C. to Fla. *quadrivittatus*.
AA. Blotched above or uniform black.
  a. Scales in 25 (rarely 23 or 27) rows; with chocolate blotches.
                                                                 vulpinus, p. 90.
  aa. Scales in 25 to 29 rows.
    Scales in 27 rows; blotches red.   *guttatus*, p. 92.
    Scales in 27 (rarely 25 or 29) rows; upper surface nearly uniform black, or grayish, with black blotches.
                                                                      obsoletus, p. 93.
    Scales in 29 rows; color ash-gray, with about 70 blotches of brown. Kansas to Mexico.   *emoryi*.

Coluber vulpinus, (B. & G.).

Fox-snake:

Scotophis vulpinus, Baird and Girard, 1853, *6*, 75; Cooper, 1860, *20*, xii, pt. ii, 299, pl. 22. *Elaphis guttatus* var. *vulpinus*, Garman, 1883, *13*, 56; *Coluber vulpinus*, Cope, 1875, *12*, 39.

Form elongated and rather slender. Tail tapering, and forming about one-fifth the total length, ending in a hard, straight spine. All the cephalic plates behind the prefrontals large. Postfrontals bent down on the sides of the face. Vertical varying from narrower than long to wider than long; shorter than distance from its anterior border to snout. Upper labials 8 (rarely 7), eye over fourth and fifth; sixth and seventh largest. Lower labials 10 or 11, the sixth largest.

Dorsal scales in 25 rows (rarely 23 or 27); the carination feeble, on the outer rows and on the tail obsolete. Ventral plates 200 to 234. Subcaudals 68 to 85.

This is a distinctly spotted serpent. The ground color above varies from gray to brown and reddish, many of the scales having a broad edge of cream color. There is a dorsal series of broad blotches of a brown or chocolate color, and these are edged with black. These blotches, about 60 from head to tail, are from three to six scales long and extend down on the sides to about the sixth row of scales. They are separated on the back by two scales' length. Alternating with these blotches is another series on each side, situated on the third to the seventh rows of scales. Below these is a third series, smaller and usually less distinct. They alternate with those of the second series, and lie opposite the dorsal spots. Sometimes they lie on a level with the second series, but are smaller. The under surface is yellowish, with large squarish blotches of black. The head sometimes with a dark streak from the eye to the corner of the mouth, and another downward from the eye.

This snake may attain a total length of five feet, and even more. Its geographical range appears to be altogether northern, from Michigan to Minnesota and south to Southern Indiana. New Harmony (Sampson's coll.); Wheatland (R. Ridgway); Hamilton County (Hay). This last mentioned specimen has 85 subcaudals.

This snake appears to be the northern representative of *C. guttatus*, a species that is at home in the Southern States, but which also is found in Indiana. By some authors, as Mr. Samuel Garman, *vulpinus* is regarded as only a variety of *guttatus*. The latter may be recognized as a redder snake, with fewer blotches along the back (about 45, instead of 60), and with 27 rows of scales.

The Fox-snake appears to be moderately common in some localities. It is often known as the "Pilot-snake," and is supposed to have some mysterious connection with the rattlesnake. It is a wholly innocent snake, although, as it seems, a little inclined to be pugnacious. Dr. Suckley (*20*, xii, pt. ii, 300) states that one of these snakes was brought to him alive at Ft. Snelling, Minn. When provoked it showed its irritation by vibrating the tip of its slender tail, which, when striking a crumpled leaf or any other small object, would produce a well-marked rattling noise, very similar to that made by the rattlesnake under the same circumstances. Other observers make mention of the same habit.

Mr. Robert Ridgway, of the Smithsonian Institution, writes me that, while hunting near Mt. Carmel, Ill., he came upon a Fox-snake over six feet in length. It immediately showed a disposition to fight, and Mr. Ridgway says that it was the most viciously pugnacious snake that he had ever seen. An examination of the stomach showed that it had just swallowed a half-grown rabbit. Its disposition appears to be in strong contrast to that of *C. obsoletus*, which, so far as I have been able to learn, is very gentle.

These snakes, being wholly harmless and subsisting on vermin of various kinds, ought to receive the protection of the farmer.

Coluber guttatus, Linn.

Spotted Coluber.

Coluber guttatus, Linnæus, 1766, *64*, ed. xii, i, 385; Holbrook, 1842, *54*, iii, 65, pl. 14; *Scotophis guttatus*, Baird and Girard, 1853, *6*, 78; *Elaphis guttatus*, Garman, 1883, *13*, 55, pl. iv, fig. 1.

Form elongated and somewhat compressed. Head narrow, tapering to the rounded snout. Tail about one-sixth the total length. Cephalic plates not greatly different from those of *C. vulpinus*. The dorsal scales disposed in 27 rows; rather feebly keeled, those of some of the outer rows smooth. Ventral plates 214 to 236; subcaudals 63 to 79.

The ground color is light red, fading in alcohol to brownish yellow. Along the back there is a series of about 40 blotches, or "saddles" (Cope), of dark red, each dark bordered. These blotches are somewhat irregular and variable in form. They are from 4 to 6 scales in length. Below this dorsal series on each side is another series of spots, alternately larger and smaller. Of these, the larger alternate with the dorsal series. All these send down prolongations to the belly. The lower surface is checkered with black and yellow. Head red, with a band of dark red, edged with black, running across the face, through the eyes, and to the corners of the mouth, and on the sides of the neck; another narrower band in front of this, and a third from the back of the head to the neck. The size is about that of *C. vulpinus*.

Distributed from Virginia to Illinois, and south to the Gulf of Mexico. It has been sent to the National Museum from Mt. Carmel, Illinois. This Mt. Carmel specimen has only twenty-six rows of scales. Some *Colubers* in the collection of Mr. A. W. Butler, of Brookville, I refer to this species. Another specimen, undoubtedly *C. guttatus*, is said to have been taken in Putnam county, at Greencastle. It is a large specimen, and has the characteristic head-bands. The scales are wholly smooth.

Of this serpent, Holbrook remarks that it is commonly observed about the roadsides early in the morning or at the dusk of evening;

unlike most snakes, concealing itself during the day. It is very gentle and familiar, frequenting the neighborhood of settlements, and at times entering houses. According to Catesby, it is a great robber of henroosts. If so, it must take the young fowls, and possibly the eggs.

Coluber obsoletus, Say.

Alleghany Black-snake.

Coluber obsoletus, Say, 1823, *14*, i, 140; Holbrook, 1842, *5*, iii, 61, pl. 12; *Coluber alleghaniensis*, Holbrook, 1842, *54*, iii, 85, pl. 19; *Scotophis alleghaniensis*, Baird and Girard, 1853, *6*, 73; *Elaphis obsoletus*, Garman, 1883, *13*, 54, pl. iv, fig. 2; *Scotophis confinus*, Baird and Girard, 1853, *6*, 76.

A snake attaining a large size, of moderate slenderness, and with a tail that forms a fifth of the total length. Head rather broad and the snout blunt. Mouth large. Eye of medium size. Crown-shields 9; those behind the prefrontals large. Postfrontals bent down on the face. Rostral broad, and the snout projecting. Upper labials 8 (rarely 9); 6th and 7th largest. Lower labials 11; the 6th largest. Scales feebly keeled; some of the exterior rows smooth; disposed in 27 rows (in some specimens, 25 rows*). Ventral plates 230 to 250; subcaudals 53 to 86.

Color varying from gray brown to pitch-black, sometimes with a tinge of red. Often with numerous evident spots; sometimes the spots obsolete, as in our form, the typical *obsoletus*. In this, the general color is a black with a bluish tinge, or a pitch-black, most pronounced on the posterior portion of the body. The anterior half may be lighter, and show evidences of blotches. The whole of this part may have a decided tinge of red, this being due to the color of the skin between the scales; yet the red may run up on the bases of the scales. Occasionally the spots of the upper surface are of a decided red. The dorsal blotches extend down on the sides to about the 7th row of scales, counting from the lowest. They are about 6 scales long, and are separated by the length of 2 scales. Alternating with these is another series which extend from the 3d to the 7th rows of scales. These spots are all feebly indicated by the sulphur yellow of the skin between the scales; and often the color is almost uniform black. There are some scales with yellow or white edges. Lower jaw and throat white. The belly is of a slate-color or black on the hinder half; anteriorly the black is mottled with yellowish, which color becomes more and more abundant, until the throat and chin are entirely yellowish. Small, or even half-grown, individuals may have a ground color of ash-gray and numerous dark blotches.

---

*A specimen from Brookville, Indiana, which resembles in other respects *C. obsoletus*, has 29 rows of scales.

Coluber obsoletus confinis is common in the southern portion of the United States. It has the scales in twenty-five, occasionally twenty-seven, rows. The ground color is ash gray, and there are about forty-four elongated dorsal blotches of dark chocolate brown. Below these blotches are two other series of elongated spots of a similar color. The latter spots run together to form on each side, especially arteriorly, a longitudinal stripe. Ventral plates about 240. A specimen of *Coluber*, fourteen inches long, sent me from Terre Haute by Prof. Blatchley, agrees in almost every respect with Baird and Girard's description of *Scotophis confinis*. It has, however, two temporal plates, instead of one, as stated by Prof. Cope. It is undoubtedly the same form as two large specimens before me, one from Georgia, the other from Mississippi. The latter belong to *Coluber spiloides*. When the small specimen is compared with other half-grown and adult specimens of *C. obsoletus*, they appear to form an unbroken series from the very spotted young up to the adults of uniform black. I conclude that *C. spiloides* is not more than a variety of *C. obsoletus*, and *C. confinis* is probably an individual variation with respect to its temporals.

Coluber obsoletus obsoletus ranges from southeast New York and the eastern base of the Alleghany Mountains to the plains, and south to North Carolina and Texas. It probably occurs in all portions of our own State. It has been taken in the following localities: Wheatland (Ridgway); Franklin county, where it is common (Hughes and Butler); Monroe county (Ind. Univ. coll.); Jackson county (St. Nor. Sch. coll.); Terre Haute (St. Nor. Sch. coll. and Prof. Blatchley); Irvington (W. P. Hay). Three of the specimens examined, one taken at Irvington, another taken in Jackson county, and a third from Terre Haute, had only twenty-five rows of scales.

So far as I am aware, this is entirely a forest-inhabiting species. Our indistinctly spotted and almost jet black form is not distinguished by most people from the Black-racer, although it is a very different snake. The latter is a slenderer snake and has very smooth scales in only seventeen rows.

Coluber obsoletus spends its time hiding about hollow logs and in holes about standing trees. It often ascends trees in search of birds and their young. Mr. Amos W. Butler, of Brookville, says that they are the most destructive to birds of all our snakes. Besides birds, they no doubt prey on mice, rats, rabbits and other small animals. The disposition of this serpent is gentle, and it makes little resistance when surprised and seized by head and tail. It will open its mouth in an attempt to bite, but struggles little. Under such circumstances a Racer would make a lively disturbance. One put into a box with a mouse would strike at the latter whenever it showed too much familiarity, but it was not harmed. In the stomach of one individual I found a number of young mice; in another were two old and six young mice.

This species probably reaches a greater size than any other snake that we have. Dr. Robert Ridgway tells me that he killed one at Mt. Carmel, Ill. which he estimated to be over nine feet long. It made no resistance when attacked and was as easily killed as an ordinary snake two or three feet long. This species has the habit, common with many snakes, of vibrating its tail so as to make a rattling or whirring sound. This probably serves to warn the larger animals of its presence, so that they may avoid it.

Dr. G. B. Goode includes this snake among those which are said to "swallow" their young; that is, when danger threatens they open their mouths, in order to allow the young to pass down the mother's throat for safety. More observations need to be made on this subject.

I have been able to find in print no observations on the breeding habits of this snake. When and where are the eggs laid? How many of these are there? How soon do they hatch? These are a few of the things that many a farmer's boy might be able to find out for us. Two individuals were taken at Fall Creek, Marion County, while in sexual union. This was on June 19. The male was 5 ft. 3 in. long; the female was 6 ft. 3 in. The female contained sixteen eggs. They have a thick covering, and must be laid before hatching. Prof. Blatchley writes (*94*, '91, 31) that he kept one, 5 ft. 7 in. long, for sometime in confinement. It would, on being disturbed, vibrate its tail in such a way as to make a rattling sound. When the room was entered at night with a lamp the snake would hiss with a loud, gurgling noise. A large Horned Owl, kept in the same room, was attacked by the snake, tightly enveloped in its coils and so badly crushed that it soon died.

Maximilian (*103*, xxxii) has confirmed the popular notion that the snake will eat fowls' eggs. One entered his room, climbed to a vessel of eggs, and swallowed a number of them. After the eggs had passed down the throat the shells were crushed by a powerful constriction of the walls of the stomach.

Genus NATRIX, Laurenti.

Natrix, Laurenti, 1768, *109*, 73; *Tropidonotus*, "Kuhl, 1826, 77, 205"; Cope, 1875, *12*, 42; Garman, 1883, *13*, 22; *Nerodia* and *Regina*, Baird and Girard, 1853, *6*, 38, 45.

Form varying from stout to slender. Head distinct from the body. Crown-shields 9. Loral present. Anteorbitals 1 or 2. Postorbitals 2 or 3. Nasals divided with the nostril between. Scales conspicuously keeled; arranged in from 19 to 33 rows. Anal plate divided.

KEY TO THE SPECIES OF *Natrix* IN THE E. U. S., NORTH OF FLORIDA.

A. Colors arranged in lengthwise stripes; scales in 19-21 rows.
   a. With a vertebral dark vitta; another on fifth row of scales; a pale stripe along the flanks, and two dark stripes on the middle of the belly. *leberis*, p. 96.
   aa. With a vertebral light band.
      b. A dark vitta on 1st row, another on 8th; two rows of black dots on middle of the belly. *rigida*, Appendix.
      bb. Dark vittæ on 4th and 8th rows; belly uniform yellow.
                                 grahamii, Appendix.
      bbb. Dark vittæ on 2d and 3d, and on 5th and 6th rows; abdomen with two clouded bands on yellow ground. Texas. *clarkii*.
AA. Colors arranged in form of blotches; scales in 19 to 33 rows.
   c. Scales in 19-21 rows. *kirtlandii*, p. 97.
   cc. Scales in 23-25 rows.
      d. Anteorbitals 1; upper labials 8. *sipedon*, p. 98.
      dd. Anteorbitals 2; upper labials 9. Washington, D. C. *bisecta*.
   ccc. Scales in 27-33 rows.
      e. Scales in 27 rows; 3 alternating rows of square spots. *rhombifera*, p. 101.
      ee. Scales in 27-29 rows; a row of small plates between the eyes and the upper labials. *cyclopion*. Appendix.
   cccc. Scales in 31-33 rows. Maryland to Florida. *taxispilota*.

Natrix leberis, (Linn.).

Leather-snake.

Coluber leberis, Linnæus, 1758, *64,* ed. x, 216; *Tropidonotus leberis,* Holbrook, 1842, *54,* iv, 49, pl. 13; Garman, 1883, *13,* 27, pl. 2, fig. 1; *Regina leberis,* Baird & Girard, 1853, *6,* 45; *Natrix leberis,* Cope, 1892, *3,* 668.

Head rather small, not much broader than the neck; somewhat depressed. Nasals 2, sometimes not well separated. Anteorbitals and postorbitals each 2. Upper labials 8, the 5th and 6th largest, the eye over 3d and 4th. Lower labials 10, the 5th and 6th largest. Scales disposed in 19 rows; all distinctly keeled, giving the snake a rough appearance. Ventrals 140 to 151. Pairs of subcaudals 64 to 86. Tail forming about one-fourth the total length.

The color above varies from olive brown to chestnut brown. There is a vertebral dark stripe occupying the median row of scales, and another on each side, occupying the 5th row from the outside. Along each flank there is a yellowish band lying on the upper half of the 1st and the lower half of the 2d rows of scales. Below this pale band there is a dark line occupying the outer ends of the ventral plates and the lower half of the 1st row of dorsal scales. The belly is yellow, with two brown bands which lie close together and run from the throat to the vent. Upper labials, lower jaw, and the throat yellow.

This species inhabits the United States east of the Mississippi River. In Indiana, it has been taken as follows: Franklin county (Butler and Hughes); Parke county (Butler); St. Paul, Decatur county (Hay); Terre Haute (Blatchley); Richmond (F. C. Test); Wabash county (W. O. Wallace). It may be expected to occur along all our streams.

This species seems to be wholly aquatic. When pursued, it exhibits great skill in secreting itself among the stones and other accumulations along the water's edge. Its colors are in a high degree protective. Little appears to be known concerning its special habits. Mr. A. W. Butler informs me that in Franklin County it appears as early as March 20, and is to be found still in November.

A gravid female 24 inches long, from Wabash county, contained 8 eggs, each of which had within it a young snake about 6.5 inches long, but not yet ready to be born. The eggs differ in form and length, on account of pressure. The two hindermost lie in the left oviduct. The egg coverings are very thin and delicate. The brown longitudinal stripes of the young are so distinct that the species might easily be determined. The brown ventral bands have not yet appeared.

Natrix kirtlandi, (Kenn.).

Kirtland's Snake.

Regina kirtlandi, Kennicott, 1856, *1*, 95; *Tropidoclonium kirtlandi*, Cope, 1860, *1*, 340; *Tropidonotus kirtlandii*, Garman, 1883, *13*, 28, pl. 1, fig. 3; *Clonophis kirtlandii*, Cope, 1888, *3*, xi, 391.

A snake having a moderately slender body, a small head, and a tail constituting about one-fourth the entire length. Head little larger than the neck. Crown-shields nine. Snout rather short. Loral present, higher than long. Nasal divided, or only partly so. Anteorbital one, high. Postorbitals two, the upper small. Upper labials six (five), the fifth the largest. Lower labials, seven or eight, the fifth and sixth the largest. Scales all distinctly keeled; arranged in nineteen rows. Ventrals 120 to 133. Subcaudals 50 to 65 pairs.

In life, the ground color of the upper surface is a dull red, most distinct along the middle line; all the scales dotted with brown. Along each flank, involving the outer one-fourth of the ventral plates and two or three of the lower rows of scales, is a band of silver-gray. There are on each side of the body three rows of alternating spots of dark-brown. Of these spots, those of the middle row are largest and most distinct. They lie on the second to the sixth rows of scales, are about two scales long, and are separated by one-half scale's length. There are about fifty-five of these spots from the head to the vent. Above this row is another of considerably smaller and fainter spots. The spots of the lowest row occupy the lower edge of each alternate scale of the first row. Belly deep salmon-red, each side ornamented by a row of distinct black spots, the size of the eye. They are located on the inner half of the outer fourth of the ventrals. Whole throat and lower jaw light salmon, as are the upper labials and the snout. Head brown, becoming black on the upper edges of the upper labials. The length may become about eighteen inches.

This species is distributed from New Jersey (*1*, '60, 340,) to Illinois. It is a common species about Indianapolis, almost as common as *Eutainia sirtalis*. It is found in Monroe county (D. S. Jordan); Crawfordsville (Beachler); Winchester (Eugle and Wright); rare about Terre Haute, common in Putnam county (Blatchley)

This is a handsome species of snake, and one that is wholly innocent. Nevertheless, it exercises the right of the innocent, and when attacked makes a show of self-defense. It has a habit of flattening itself excessively, so that it becomes very broad and thin. It will strike vigorously, but does no harm. It appears early in the spring, and is seen late in the autumn. In a mild winter and on a sunny day, I have seen it on January 25. On the other hand, I have seen it as late as the middle of October; indeed, it appears to be more abundant late in the autumn than in the summer. About this date, several half-grown ones were found. On the 21st of March, one was dug up out of the mud on the margin of a pond. A specimen from Winchester, Indiana, contained eight eggs. These had, apparently, not begun development, and were only about .44 inch long. The species probably produces living young.

Natrix sipedon, (Linn.).

Water-snake.

Coluber sipedon, Linnæus, 1758, *64*, ed. x, 219; *Tropidonotus sipedon*, Holbrook, 1842, *54*, iv, 29, pl. 6; Garman, 1883, *13*, 25, pl. 2, fig. 3; *Nerodia sipedon*, Baird and Girard, 1853, *6*, 38; *Natrix fasciata sipedon*, Cope, 1892, *3*, 671.

Variety *fasciata*. *Tropidonotus fasciatus*, Holbrook, 1842, *54*, iv, 25, pl. 5; *Coluber fasciatus*, Linnæus, 1766, *64*, ed. xii, i, 378; *Nerodia fasciata*, Baird and Girard, 1853, *6*, 39; *Natrix fasciata*, Cope, 1892, *3*, 670.

Variety *erythrogaster*. *Coluber erythrogaster*, Shaw, 1804, 71, iii, 458; *Tropidonotus erythrogaster*, Holbrook, 1842, 54, iv, 33, pl. 7; *Nerodia erythrogaster*, Baird and Girard, 1853, 6, 40; *Natrix fasciata erythrogaster*, Cope, 1892, 3, 673.

Variety *woodhouseii*. *Nerodia woodhouseii*, Baird and Girard, 1853, 6, 42.

This widely distributed and variable species has been described under a great number of names, each of which has been regarded as that of a distinct species. It is doubtful if some of the above will be able to maintain their position even as recognized varieties.

Natrix sipedon is our most common Water-snake, and when fully grown attains a large size. The head is rather narrow, and pointed in front. Anteorbital 1, high. Postorbitals 3. Nasal sometimes divided above. Upper labials usually 8, the 6th and 7th large, the eye over 4th and 5th. Inferior labials 10. Scales strongly keeled; arranged in 23, rarely 25, rows. Ventral plates 135 to 150. Subcaudals 60 to 75.

The ground color varies from ashy to brown. Along the back there is a series of brown spots, about 30 in number, from the head to the vent. These are about three scales long and descend on the sides to the sixth row of scales. They are separated on the back by the length of one-half scale. Along each side is another series of similarly colored square blotches. These either come opposite the dorsal blotches or alternate with them. When the three series are opposed, they unite to form continuous cross bands over the body. Spaces between the lateral blotches equal to or less than width of blotches. When they alternate, they touch at their contiguous angles. All the blotches are dark edged, while the ground color around each is paler. Occasionally the blotches unite across the back to form oblique bands. Again, the ground color is sometimes so dark that the blotches are obscured. Upper surface of the head brown. Upper labials, lower jaw, throat, and sometimes the anterior portion of the belly, yellow. The brown of the belly often takes the form of triangular spots and these are often suffused with red. On the outer ends of the ventral plates anteriorly there begin to be some mottlings of brown. Further back this increases, until the whole or nearly the whole, of the belly is brown.

In the young the contrast between the ground color and spots is greater, the spots being nearly black.

This variety of the species *N. sipedon* is distributed over the country from the Atlantic to the Mississippi Valley, and south to the Gulf. It is the common Water-snake of Indiana and it seems to be needless to specify localities.

Natrix sipedon fasciata. This form of *sipedon* is characterized by having a series of dorsal blotches of brown or black, about thirty in number to the vent, and these spots run down on the sides of the animal, becoming narrower. The spaces between the blotches are occupied on the flanks by red spots, which extend down to the ventral plates. Sometimes the red spots lie just below the dark dorsal blotches, and are separated by narrow dusky bands that connect above with the dorsal blotches. Underneath the color is reddish white, and there may be some dark marbling.

This variety is abundant in the Southern States, and has been taken by Mr. Robert Ridgway at Wheatland.

N. sipedon erythrogaster. In this variety the upper surface is almost or entirely without dark blotches, the color being a uniform blue-black. The head and neck may be almost black. The belly, as its name implies, is more or less red in life, yellow in alcohol. Holbrook's figure of this snake represents its under surface as being of a coppery red. It, too, is found mostly in the South, but has been reported from Southern Illinois, and from Mt. Carmel, on the Wabash River. A specimen from that place is in the National Museum. It may, therefore, be regarded as an Indiana serpent. Mr. F. C. Test, of the National Museum, tells me that this snake is found in Hamilton county. Mr. Robert Ridgway says that it is very abundant at some points along the Wabash River. Mr. A. B. Ulrey, of Wabash county, has shown me a large specimen which was obsoletely blotched above, red below, with some brown mottlings. It appears to be intermediate between *sipedon* and *erythrogaster.*

N. sipedon woodhousei. This variety has not yet been reported from Indiana, but it is in the National Museum from St. Louis, Mo., and even from Northern Illinois. Its range is toward Texas. It is distinguished by having the scales in twenty-five rows. The lateral blotches alternate with the dorsal and are elongated downward, and are separated by spaces wider than themselves. It deserves to be sought for in our State.

Natrix sipedon, known as the "Water-snake," "Water Moccasin," is extremely abundant in all our streams. Under the impression that it is poisonous it is greatly feared by many people, who suppose it to be the same as the poisonous moccasin of the Southern rivers. This is a mistake, and the name moccasin ought not to be applied to our species. It has no poison fangs whatever, and its bite would produce nothing more serious than a fright and a few scratches. This snake is, however, of an ugly and sullen disposition, and when caught will struggle and strike and bite. It may be seen along the river banks, gliding from stone to stone, or swimming hastily away, to escape observation. When pursued it will dive to the bottom and conceal itself among stones and vegetation. Holbrook states that it is frequently seen resting on the low branches of trees that overhang the water. Of *fasciata,* whose habits are doubtless identical with those of our form, Holbrook says that it is a bold animal, and is one of the very few snakes that will, in confinement, devour its prey. These serpents all probably leave the water at times during the night and wander about. I have a large specimen of *erythrogaster* that, in the night, fell into a well that was being dug.

The food of this snake consists of frogs, fishes, and similar animals. There is an account in *Nature* of one which was found swallowing another. Prof. Blatchley (*94*, '91, 30) says that he found seven large Leopard Frogs in the stomach of one of these snakes. Dr. J. A. Allen tells of seeing one brought from the water with a pickerel a foot long in its mouth.

These snakes are ovoviviparious, that is, they retain the eggs in the body until they are hatched. In some cases the eggs may be laid a little before the hatching takes place. In *1*, 1887, 121, we are told of a case in which 33 young were taken from the body of one large female. Each one had attached to it a portion of an egg, from which it was absorbing nutriment. Prof. F. W. Putnam tells us of a family of 22 of the young of this snake, which were found in Massachusetts. Each one was 8 inches long. Another smaller specimen from Northern Indiana contained 16 eggs. The young in these eggs were 7.5 inches long. Each of these was provided with a well-developed egg-tooth.

While investigating the question whether or not snakes "swallow" their young, Prof. G. B. Goode found evidences that this act had been observed in the case of the Water-snake seventeen different times. This being such a common snake further observations ought to be made. If one thinks that the young have been seen to enter the mother's mouth she ought, if possible, to be caught, handled carefully, and put into a cage, to see if the young will come out again. Or, if she is killed, a careful dissection ought to be made, in order to learn whether or not the young are really in the stomach.

Natrix rhombifera, (Hallowell).

Diamond Water-snake.

Tropidonotus rhombifer, Hallowell, 1852, *1*, 177; *Nerodia rhombifer*, Baird and Girard, 1853, *6*, 147; *Natrix rhombifera*, Cope, 1892, *3*, 673.

Head narrow. Anteorbital 1, sometimes 2. Postorbitals 3, the lowest nearly meeting the anteorbital under the eye. Upper labials 8, the sixth and seventh largest; the eye over third and fourth, not in contact with the fourth on account of the lowest postorbital. Lower labials 11, the fifth and sixth largest. Scales in 27 rows, all conspicuously keeled. Ventrals 136 to 141. Subcaudals 62 to 70.

Ground color above reddish gray. On the middle of the back there is a series of about 50 squarish brown blotches. Alternating with the dorsal series there is, on each side, a series of similarly colored blotches. The dorsal spots are about two scales long, and separated by the length of three scales. The lateral blotches reach down to the ventrals and even lower; while above they join the contiguous corners of the dorsal blotches. Of these blotches about 32 lie in front of the vent. Occasionally there is a little confusion in the relations of the blotches of the different series, but not much. The belly is yellowish white, with a few triangular spots of black, giving it a speckled appearance. The head is smoky brown above as far down as the upper edges of the upper labials. The lower edges of these are yellow, with a black border on the hinder edge. The lower labials are similarly yellow, with black posterior edges.

The size is about that of *Natrix sipedon*, 2 feet 8 inches or more.

This species is distributed from Michigan to Louisiana and Texas. There is a specimen of it in the National Museum from Lafayette Others have been sent there by Mr. Robert Ridgway from Wheatland. I have seen another specimen in Mr. Sampson's collection made at New Harmony. It is no doubt generally distributed throughout the State.

Of the habits of this species I know nothing, except that it is an aquatic snake.

Genus HETERODON, Beauv.

Heterodon, Pal. de Beauvais, 1802, 57, iv, 32; Holbrook, 1842, 54, iii, 37; Baird & Girard, 1853, 6, 51.

Body short and stout. Neck nearly as thick as the head. Tail short. Head broad, short and high. Head, neck and body capable of great flattening. Outline of mouth much curved, rising behind the eyes. Posterior teeth longer and fang-like. Snout projecting beyond the mouth, shovel shaped. The rostral with a sharp horizontal edge and a longitudinal ridge above. This followed behind by a small azygous plate. Upper labials cut off from contact with the eyes by suborbital plates. Dorsal scales keeled. Anal plate divided.

KEY TO THE SPECIES OF *Heterodon*.

Prefrontals separated by the azygos only.   *H. platirhinos*, p. 102.
Prefrontals and sometimes postfrontals separated by a number of small plates.   *H. simus*, p. 105.

Heterodon platirhinos, Latr.

Hog-nosed Snake; Spreading Viper.

Heterodon platirhinos, Latreille, 1802, 57, iv, 32; Holbrook, 1842, 54, iv, 67, pl. 17; *H. platyrhinos*, Baird & Girard, 1853, 6, 51; Garman, 1883, 13, 75, pl. 6, fig. 5; *Heterodon niger*, Troost, 1836, 62, iii, 186; Baird & Girard, 1853, 6, 55.

Body stout and heavy. Tail about one-fourth the entire length, often shorter. Head rather broad and short, but quite deep. The usual nine crown-shields present. In addition, there is an azygous plate between the prefrontals. Rostral greatly developed, trihedral, pointed, and upturned. Nasals two, with the nostril between. Lorals two, these sometimes united. The anteorbitals forming a part of the row of small plates which encircle the eye. This ring is completed above by the supraoccipital. Mouth-cleft large and much curved. Upper labials eight, lower labials eleven. Eye large. Scales keeled, except those of the outer row, which are smooth; arranged in twenty-five, occasionally twenty-three, rows. Ventral plates 120 to 150. Subcaudals 45 to 60.

The color of the upper surface varies from yellowish, reddish brown, and even brick red, with dark blotches, to a uniform gray, olive-brown, blue-black, or black. When the color is not uniform there are to be seen three series of dark spots. One of these is vertebral, and consists of from twenty to thirty brown blotches, each from two to five scales long. They extend down some distance on the sides, but are likely to fade out into the ground color. On the median line, these blotches are separated by about the length of three scales. Alternating with the dorsal blotches is a series of lateral spots, circular and of a dark color. These are not developed on the tail, while the dorsal blotches form bands that extend nearly around the tail. Sometimes there are small spots on the lowest rows of scales forming a third series. All the spots are variable in size and depth of color. Head with a dark band running across from one orbit to the other, a black band from the eye to the corner of the mouth, and another across the occipital plates on to the sides of the neck. The under surface is yellow, with some mottlings of brown.

The uniformly colored individuals which were for a long time regarded as a distinct species, differ in having the color uniform gray to black above, slate color below. There are, however, all gradations in the amount of black and the distinctness of the spots to be found, and these completely connect the two supposed species.

This snake may attain the length of three feet. The body being heavy, the size then becomes conspicuous.

The distribution is from Pennsylvania to Florida and west to Minnesota and Texas. I have reports and specimens from many points in Indiana. Both spotted and uniform specimens are found at New Harmony (Sampson's coll.); also about Brookville (Hughes). I have specimens from Vernon, Jennings county (J. Cope); from Fountain county (C. H. Smith, Veedersburg); Vigo and Jackson counties (Normal School coll.); said by Mr. Beachler to occur at Crawfordsville, Montgomery county; Brown county (Chas. Jameson). Specimens sent me from Veedersburg by Mr. Smith are intermediate between the two supposed forms, and the same is true of some in the collection made by Prof. Evermann, at Terre Haute. Prof. Blatchley reports having found the two forms copulating together (*94*, 1891, 32). Mr. Quick, of Brookville, tells of a similar case.

This is a short-bodied and clumsy snake. It has a peculiar habit, when it is disturbed, of flattening out its head and anterior part of the body, so that it has quite a formidable appearance. It is said to inflate the skin of the head and neck, but it is doubtful if it really does this. When enraged it hisses loudly, and this has given origin to the name, "blowing adder." The snake has a very bad reputation among the people, as is shown by the names "adder" and "viper." Long ago a writer said of it: "When approached it becomes flat, appears of different colors, and opens its mouth hissing. Great caution is necessary not to enter the atmosphere which surrounds it. It decomposes the air, which imprudently inhaled induces languor. The person wastes away, the lungs are affected, and in the course of four months he dies of consumption." The popular notion about this snake even yet is that it is very poisonous, and that it can even spit venom. This bad reputation may be due to its resemblance to some of the poisonous snakes, and to the great show of bravery which the snake often indulges in. It is, on the contrary, regarded by herpetologists as one of the most harmless of our serpents. At the base of its upper jaws there is on each side a long tooth, which some have supposed might act as a poison fang. It has no canal or groove, is not connected with any known source of poison, and is, moreover, so far back that it is hard to see how it could be used in striking an enemy. It would be of immense advantage in swallowing frogs. A number of scientific men have reported that they have allowed themselves to be bitten by this snake and have received no harm. When attacked it spreads out its head and body, utters a hiss, and presents a threatening appearance. However, when it is tormented it will feign death, sometimes turning over on its back and remaining motionless, and will repeat this action.

Prof. W. S. Blatchley says of two, a black one and a spotted one, which he disturbed, "that they opened wide their mouths, turned on their backs and coiled and twisted about in a very rapid and curious manner for about five minutes, when they became quiet and apparently lifeless. During all these contortions they had remained on their backs, and when they became quiet and were turned over they would immediately turn on their backs again, but otherwise gave no signs of life, even at the end of an hour's time."

This species is said to prefer dry and dusty fields; but Dr. C. C. Abbott found them in spring along ditch banks looking out for Cricket-frogs, which a dissection showed that the snakes had been eating. Much appears to be unknown about the habits of this snake, as of most others. What is the use of the sharp-edged, pointed snout? Is it used in burrowing? Or is it employed in rooting up the ground in search of insects and other food? Again, what influence does cultivation of the soil have on the numbers of such sluggish serpents? Why do these survive, while rattlesnakes and copperheads so rapidly disappear? The food of these snakes is doubtless principally frogs, toads and probably mice. Rev. Samuel Lockwood tells us (*22*, 1875, 10) that he has known them to eat the heads of the common eel left on the shore by fishermen.

Some facts are known about its breeding habits. Troost dissected a specimen and found in her 25 oval eggs, each three-fourths inch long and without a calcareous covering. In a large female of the common form from Veedersburg I found in one oviduct 4 eggs, and in the other 11. The hindermost egg was an inch and a quarter long and three-quarters across. The eggs were covered with a tough membrane. I found no embryos in any of the eggs. The snake has been regarded as ovoviviparous, but such is probably not usually, if ever, the case. We have evidences that the eggs are usually laid and buried in the earth before they are ready to hatch. Prof. F. W. Cragin (*22*, xiii, 710) says that he had 22 eggs of this snake, which had been plowed up in a sandy field along Long Island Sound. One of these hatched four days afterward. Another writer (*22*, iii, 555) states that he saw one of these snakes killed, and out of a wound in its side there issued over 100 young, each about 6 or 8 inches long. This writer believed that these snakes were ovoviviparous, and that these young had really been in the stomach of the female The number of young in this case is certainly unusual. Another author states that a "spotted spreading adder" contained 87 young, each nearly 6 inches long. These statements about such large numbers of eggs are undoubtedly erroneous. A nest of 27 eggs was brought to the National Museum August 31. The female was near the nest and attempted to defend it. In each of the eggs was an embryo well developed and about 8 inches long. The eggs did not hatch until September 7 and 8. The egg coverings were ripped open by the egg-tooth of the young snake. The young would flatten themselves when teased, and some would feign death. Dr. G. B. Goode states (*34*, 1873, 184) that the female of this species has been reported as affording its young a hiding place in her stomach.

Heterodon simus, (Linn.).

Sand Viper.

Coluber simus, Linnæus, 1766, *64*, ed. xii, i, 216; *Heterodon simus*, Holbrook, 1842, *54*, iv, 57, pl. 15; Baird and Girard, 1853, *6*, 59; Garman, 1883, *13*, 76, pl. vi, fig. 4.

Form much that of *H. platirhinos*, but probably not attaining so large a size. Tail shorter, about a sixth or less of the total length. Head short and broad. All the crown-shields short and broad. Mouth-cleft much curved. The snout upturned. Prefrontals and postfrontals separated by a number of small plates, which surround the azygos. Upper labials, 8, the sixth very high, but all excluded from contact with the eye by the suborbitals. Lower labials 9 to 11, small. Scales not conspicuously keeled, disposed in 25 or 27 rows. Ventrals 117 to 150. Suborbitals 32 to 55.

The ground color is olive or yellowish brown, with many scales partially or wholly yellowish. The upper surface is relieved by a dorsal series of brown or blackish blotches, and two lateral series on each side. The spots lowest down on each side are almost obsolete. The median series consists of about 35 somewhat irregular blotches, each 3 or 4 scales long. Alternating with these is a series of round spots about 3 scales in length. All the spots are surrounded with an edging of yellow. There is a narrow black band running across the forehead, through the eyes and to the corner of the mouth. The belly is yellow, with some cloudings of brown.

The length of this species is not so great as that of *H. platirhinos*, probably never exceeding two feet.

Heterodon simus simus is found from South Carolina to Mississippi, and north to Indiana. Further west it is replaced by *H. simus nasicus*, a variety with a still more prominent rostrum, about 50 spots in the dorsal series, a greater number of small scales around the azygos, 23 rows of scales, and more black on the belly.

In Indiana *simus* has been taken at very few points, but it is probably to be found throughout at least the southern portion of the State. It has been sent to the National Museum from Brookville by Dr. R. Haymond. It has been stated to occur at New Harmony, but I have not seen specimens from that place.

I find few observations that have been made on the habits of this species. The form *nasicus*, being more abundant, has been a little better observed, and its habits probably do not differ much from our variety. It is much dreaded by the settlers in the West, who call it the "Sand Viper." It is entirely harmless, and Dr. Coues states that he could not provoke it to bite. Like its relative, *platirhinos*, when it is disturbed it makes a great demonstration of hostility. One that I came upon in a prairie path in Kansas sprang out into plain sight and began such a wriggling and hissing that for a moment I supposed that it was really a rattlesnake. This was probably its purpose, and such movements would be quite likely to frighten away from it any large animal that otherwise might trample it.

One writer (22, xvi, 566) tells of having found this snake hanging to the foot of a box-tortoise. The foot was bleeding, and two of the toes had been digested off. These tortoises are not infrequently found with parts of the limbs missing, and this observation may account for such maimings.

Genus OPHIBOLUS, B. & G.

Ophibolus, Baird and Girard, 1853, *6*, 82; Garman, 1883, *13*, 64.

Form moderately stout. Head not much broader than the neck. Crown-shields 9. Loral plates present, or only rarely absent. Anteorbital single. Postorbitals 2 or 3. Nasals divided, with the nostril between. Scales smooth and shining, disposed in 21 to 25 rows. Anal plate entire.

KEY TO THE EASTERN UNITED STATES SPECIES OF *Ophibolus*.

A. Scales in 21 (rarely 19 or 23) rows.
   a. General color red, with 22 or more triple rings of black, red, and black; or ground color reddish or gray, with 40 to 50 chocolate or brown saddle-shaped blotches.
                                                                doliatus, p. 107.
   aa. Ground color chestnut or olive, with about 50 rhombic blotches along the back. Other blotches along the sides. Southern States, north to District of Columbia.
                                                                rhombomaculatus.
   aaa. Ground color black, with numerous small spots of yellowish or white, these sometimes forming transverse rows of spots.
                                                                getulus, p. 101.
AA. Scales in 25 rows.
   Brownish, with 40 to 60 dark blotches above. Belly blotched.
                                                           calligaster. Appendix.

Ophibolus doliatus, (Linn).

House Snake; Milk Snake; Chicken Snake

Coluber doliatus, Linnæus, 1766, *64*, ed. xii, 379; *Coronella doliata*, Holbrook, 1842, *54*, iii, 105, pl. 24; Baird and Girard, 1853, *6*, 89; Garman, 1883, *13*, 64 and 155.

Variety *triangulus*. *Coluber triangulus*, Boie, 1827, *77*, 537; *Coluber eximius*, Holbrook, 1842, *54*, iii, 69, pl. 15; Baird and Girard, 1853, *6*, 87; *Ophibolus doliatus*, var. *triangulus*, Cope, 1875, *12*, 37; Garman, 1883, *13*, 65, pl. 5, fig. 1.

Variety *syspilus*. *Ophibolus doliatus syspilus*, Cope, 1888, *3*, 384.

Ophibilus doliatus is an extremely variable species, and this has caused it to be described under a great number of names, the most of which have at one time or another stood for distinct species. Prof. Cope (*3*, xiv, 608), in a general view of the species, recognizes no fewer than 11 varieties or subspecies. In its distribution, the species ranges from Maine to Panama, and illustrates well the changes which a widely spread species undergoes, when traced from one region to another. The changes concern its size, squamation, its colors and the arrangement of these colors.

The head is small, little exceeding the diameter of the neck. The loral is present, with rare exceptions, but it is small. Upper labials 7; lower labials 9, occasionally 10. The eye is small and placed over the third and fourth labials. The scales are in 21 rows. In a few cases there may be only 19, and in rarer cases 21 rows. They are smooth. The ventrals range from 184 to 214. The tail forms one-sixth or one-seventh of the entire length. The ground color varies from red to reddish brown and to gray. The deeper colors appear as transverse rings, complete or incomplete, or as saddle-shaped and irregular blotches. In *O. d. coccineus*, a Southern variety, the color is red, and there are triple rings entirely encircling the body. These consist of two rings of black enclosing between them a ring of white or yellow. In other varieties the black rings of one set runs forward on the sides and joins the posterior ring of the set in front. In this way the ground color becomes divided up into a number of dorsal blotches or "saddles." At the same time the hue of both the ground color and of those which relieve it undergo change.

KEY TO INDIANA VARIETIES OF *O. doliatus*.

General color red; large blotches, or saddles of ground color formed by rings of adjacent sets meeting on the ventral plates. *doliatus*.

General color red; blotches of ground color formed by the black rings of adjacent sets meeting above the ends of the ventrals; single series of black spots along middle of belly. *syspilus*.

General color brownish-red or gray, with dorsal saddles not descending to the ventral plates. *triangulus*.

Ophibolus doliatus doliatus is bright red above, yellowish in alcohol. Across the body there run from 20 to 30 triple rings, of which two black ones enclose one of white or yellow. The black rings do not pass entirely around the belly, but the anterior black ring of one set turns forward on the ends of the ventral plate on each side and joins the posterior black ring of the set in front. On the sides are spots alternating with the dorsal blotches, and these do not meet on the belly.

This variety probably does not attain a size greater than about two feet. It is found principally in the Southern States, but has been taken also in Indiana. I have seen specimens taken at New Harmony (Sampson's coll.); Brown county (taken by Chas. Jameson); and Jackson county (Nor. Sch. coll.). The latter specimen had 23 rows of scales, otherwise it was normal. Mr. Robt. Ridgway writes that it is not uncommon about Wheatland; Terre Haute (Blatchley); Lafayette (F. C. Test).

Ophibolus doliatus syspilus is a variety described in 1888 by Prof. E. D. Cope (*3*, xi, 384). It differs from the variety *doliatus* in having the dorsal blotches descend upon the ventral plates, and in having the spots which alternate with them meet on the belly and form there on the middle line a single series. "The body is scarlet, banded with 22 pairs of jet black rings, with a white ring between each pair of black. * * The belly is marked with a single series of median black spots, which are opposite the spaces between the dorsal saddles, or opposite the yellow rings." The length of the longest specimen is 30.5 in. Most of the specimens upon which this species are founded are from the Southern States, but one was taken at Wheatland, Ind., by Mr. Robert Ridgway.

Ophibolus doliatus triangulus is by far the commonest variety of its species that is found in Indiana. It is known to all under the familiar names of "House-snake," "Milk-snake," and "Spotted Adder." The ground color above is ashy, more or less mingled with brown, this often in life tinged with red. Above are five series of blotches and spots, a dorsal, and on each side two lateral. The spots of the dorsal row are the largest, broader than long, and extend down on the sides, but do not reach to the ventral plates. They vary in number from 40 to 60. Their color varies from grayish brown to brownish red, and are more or less distinctly bordered with black. This black sometimes becomes so conspicuous as to give the appearance of half rings on the upper surface. The dorsal blotches often coalesce with the spots of the series below. These latter are small and alternate with those of the dorsal series. They are sometimes missing. The blotches of the lowest series are located principally on the ends of the ventral plates, and usually coalesce with the spots of either the dorsal or of the middle series. They are commonly black or blue black. The anterior blotch usually extends forward on to the head and terminates variously. There is commonly a dark band across the top of the head just in front of the eyes, and this runs back through the eyes on the lower neck. Above it is a yellow band that also extends backward on the neck. At the back of the head, surrounded by the termination of the first dorsal spot, is a spot of yellow. The belly is blotched with black and creamy white, the latter in life tinged with red.

The length of this form may become as much as four feet. It is common throughout the State.

DeKay (*30*, 38) says of this snake that it is found not infrequently in outhouses and in dairies or cellars where milk is kept, which it is said to seek with avidity. It climbs well, and glides rapidly over the smoothest surfaces. He adds that its vivid colors change almost immediately after death. Holbrook states that this serpent is gentle in its habits, feeding on mice, various insects, etc. It approaches without fear the habitations of men, and is, hence, not infrequently called the House-snake. It also frequents dairies where milk is kept, and this, from a mistaken notion of its robbing the dairy-women, has given rise to another name, the Milk-snake. A number of snakes have acquired a reputation of drinking milk, and it would be interesting to have some one settle the question whether or not any of our snakes will really do so. This snake appears early in the spring. I have taken it March 20, and have seen a dead one along the road still earlier.

According to Dr. Goode's investigations this snake is oviparous and guards its nest. Moreover, when danger threatens her young, the mother finds an asylum for them down her capacious throat. (See *34*, '73, 182.) I have taken the eggs of this species in Illinois. They were buried in a pile of manure and more or less glued together. The egg is 2 inches long and a little less than 1¼ inches short diameter. The covering is parchment-like. It contained a young snake 10 inches long.

Dr. C. Hart Merriam (*78*, '78, 69) relates an instance of an individual of this species having swallowed a specimen of a Striped-snake two-thirds its own length.

Ophibolus getulus, (Linn.).

King-snake.

Coluber getulus, Linnæus, 1766, *64*, ed. xii, i, 382; *Coronella getula*, Holbrook, 1842, *54*, iii, 95, pl. 21; *Ophibolus getulus*, Baird and Girard, 1853, *6*, 85; Garman, 1883, *13*, 68, pl. 5, fig. 3.

Variety *sayi*.

Coronella sayi, Holbrook, 1842, *54*, 99, pl. 22; *Ophibolus sayi*, Baird and Girard, 1853, *6*, 84; *Ophibolus getulus* var. *sayi*, Cope, 1875, *12*, 37; Garman, 1883, *13*, 68, pl. 5, fig. 4.

Ophibolus doliatus var. *sayi* has a rather elongated and slender body and small head. The tail forms only about an eight of the total length. The vertical plate is triangular. Anteorbital single; postorbitals two, small. The loral is small. The snout is pointed and projecting. Upper labials 7, the eye over the 3d and 4th. Lower labials 10, the 4th and 5th largest. Ventral plates 210 to 225; subcaudals 48 to 55.

The color above is deep brown to jet black, more or fewer of the scales being ornamented each with a small spot of white or yellow. These spots usually largest on the lowest scales. Often they are scattered over the body quite regularly; at other times there are rows of these spots which form narrow lines across the back at intervals of about 5 or 6 scales. In this case there may be as many as 70 of these cross-lines. The areas between such lines may be without spots or may be spotted. Sometimes the lines fork on the sides, the branches anastomosing so as to divide the ground color into more or less distinct blotches. Each plate of the head with one or more spots of yellow. Labials bright yellow, bordered with black. Throat yellow. Belly yellowish, checkered with blue-black, the blotches following the plates. The length may be four feet or more.

This variety is distributed from Southern Indiana and Illinois south and southwest to Mexico. A specimen found in Mr. Sampson's collection at New Harmony and assigned by me to the typical *getulus* (*94*, 1887, 64) is probably more closely related to *sayi*. *Sayi* has also been taken at Mt. Carmel, Ill., on the Wabash River. Prof. Blatchley reports having found a small specimen at Terre Haute. Recently Mr. F. C. Test of the U. S. Nat. Museum has shown me a large specimen of the variety *sayi* that he took at Lafayette, Ind. This is by far the most northern locality for the State, if not for the Mississippi Valley. The cross-bands of white are very narrow, and there are few spots on the scales between the bands. Length 3ft. 2 in.

Ophibolus getulus getulus differs from *sayi* in having the white and yellow more definitely disposed in bands which divide up the ground color into blotches. Of these there are a dorsal and a lateral series on each side. There are about 25 to 35 of the yellow cross-bands over the back. Usually the lateral blotches alternate with the dorsal. Occasionally a specimen may have more yellow than black. The belly has various proportions of the yellow and the black. Southern in distribution, but likely to occur in Indiana.

Ophibolus getulus niger is a so-called variety of this species which Yarrow (*3*, v, 438) has described from specimens sent to him from Wheatland by Mr. Robert Ridgway.

It is characterized by the almost complete absence of the usual yellow spots. The head plates are entirely black, not spotted. On the body there are spots on only a few of the scales of the lower rows. The abdomen is dull black, with white blotches. One specimen taken at Wheatland was 4 feet 6 inches long. This form is reported to be common in that region. We have in the King-snake another illustration of the phenomena seen in *Heteroden platirhinos*, *Natrix sipedon*, and some others, a gradual transition from very spotted specimens to those which are of a plain black color. We can not give a distinct name to every stage in the change.

Habits.—The habits of the different varieties of this species are, in all probability, identical. It appears to be a snake of a mild and harmless disposition. It is extremely active and strong. Holbrook says of it that it is found abundantly in moist and shady places, although it never takes to water or trees. It feeds on moles, small birds, or such reptiles as lizards, salamanders, toads and the like, that fall in its way. He further says that it is commonly believed that it is the great enemy of the rattlesnake, but that there is no great evidence of this. He, however, tells of one that had as a fellow-prisoner a *Crotalophorus miliarius*, or Southern Ground Rattlesnake, and swallowed him. I found that in Mississippi this snake had the reputation of destroying rattlesnakes, and it received protection on this account. Dr. Elliott Coues (*9*, 4, 269) says that the Black-snake (*Bascanion constrictor*) and *Ophibolus getulus sayi* wage a constant warfare against rattlesnakes and moccasins. They are said to be uniformly victorious and to eat their victims. It is on account of their prowess in thus destroying poisonous serpents that they have received the name of King-snake. Mr. J. T. Humphreys, Burke county, North Carolina, gives (*22*, xv, 561) an interesting account of a conflict in a cage between a king-snake, *sayi*, and a water-moccasin. The former was 42 inches long, the latter 34, but with a considerably larger body. The moccasin was killed, its bones crushed, and, beginning at the head, the king-snake swallowed 16 inches of the moccasin's body. Chloroform was then administered and both snakes preserved. The king-snake had previously, while in captivity, eaten seven snakes. Dr. Yarrow (*22*, xii, 470) describes a specimen of *getulus* in the National Museum that has two perfect heads. One head is a little larger than the other. The two gullets unite to pass into the one stomach.

Professor Cope (*3*, xiv, 613) states that the variety *getulus* is entirely inoffensive to man, making no hostile demonstrations. His daughter, when a girl of 6 or 8 years, had several individuals as pets. They drank milk readily from a cup which she held in her hand.

Genus EUTAINIA, B. & G.

Eutainia, Baird and Girard, 1853, *6*, 24; *Eutænia*, Cope, 1875, *12*, 40.

Form ranging from stout to slender. Head separated from body by an evident neck. Crown-shields 9. Loral present. Nasals divided, with the nostril between. Anteorbital 1. Postorbitals 3. Anal plate entire. Scales keeled, arranged in 19 to 21 rows.

A genus of snakes closely allied to *Natrix*, but differing in the undivided anal plate. Confined to North America, consisting of about a dozen species, all intimately related, and some of them exceedingly variable.

The general color consists of three light stripes on a darker ground, the intervals with alternating or tesselated spots. Abdomen without square spots. (Baird.)

KEY TO THE INDIANA SPECIES OF *Eutainia*.

A. Scales in 19 rows.
   a. Lateral stripe covering two rows of scales.
     b. Lateral stripe on third and fourth rows of scales.
       c. Very slender. Tail about one-third the total length.
          saurita, p. 113.
       cc. Stouter. Tail one-fourth or less of the total.
          radix, p. 115.
     bb. Lateral stripe on second and third rows of scales.
          sirtalis, p. 117.
   aa. Lateral stripe covering three rows, second, third and fourth.
          butlerii, p. 120.
AA. Scales in twenty-one rows (sometimes).   *radix*, p. 113.

Eutainia saurita, (Linn.).

Ribbon Snake.

Coluber saurita, Linnæus, 1766, *64*, ed. xii, i, 385; *Leptophis sauritus*, Holbrook, 1842, *54*, iii, 21, pl. 4; *Eutainia saurita*, Baird and Girard, 1853, *6*, 24; *E. faireyi* and *E. proxima*, Baird and Girard, 1853, *6*, 25; *Tropidonotus sauritus*, Garman, 1883, *13*, 23, pl. 3. fig. 2.

An elongated species, with a slender neck, and a tail that forms from a little more to a little less than a third the total length. Crown-shields normal. Loral elongated. Nasal distinctly divided, with the nostril between them. Anteorbital 1, high. Upper labials 7 or 8. Lower labials 10, fifth and sixth large. Scales distinctly keeled, and disposed in 19 rows. Ventrals 150 to 180. Subcaudals 100 to 120.

The color above consists of a light chocolate to black ground, relieved by three stripes of greenish white or yellow. The dorsal stripe lies on the median row of scales and the adjacent half of the next row on each side. The lateral stripes lie on the third and fourth rows above the belly. The abdominal surface is greenish white, without markings or spots of any kind.

Three varieties of this species are here recognized, *saurita*, *faireyi* and *proxima*. These are still regarded by most American writers as distinct species, but I have been unable to find any characters by which the many connecting forms may be satisfactorily distributed. Prof Cope, in his most recent publication (*3*, xiv, 646, 650), unites *faireyi* with *proxima*, and distinguishes these from *saurita* by their having 8 upper labials instead of 7. It is doubtful if this will hold good.

Variety *saurita*.

This is distinguished by having the tail, on an average, a little more than a third of the total length; ventral plates from 150 to 170; color light chocolate to deep brown, with the stripes uniform yellow. The slenderest of all the varieties. Atlantic to Mississippi Valley.

Variety *faireyi*.

Stouter in form. Tail rather less than a third the total length. Ventral plates 168 to 180; color blackish brown, with the stripes uniform greenish yellow. Wisconsin to Louisiana.

Variety *proxima*.

Stoutest in form, the tail being contained in the total length about three and a half times; color black, with the dorsal stripe brownish yellow, the lateral greenish; ventrals 170 to 180. Wisconsin to Louisiana and Texas.

Specimens sent to Professor Cope by Mr. A. W. Butler, and taken probably at Brookville, have been pronounced *Eutainia saurita*. Others of this variety have been secured at Wheatland by Mr. Robert Ridgway, and in Brown county by Mr. Charles Jameson. I have specimens from Fountain county, sent by C. H. Smith, and there is record of a specimen in the National Museum from Boone county.

Of *faireyi* I found a specimen in Mr. Sampson's collection at New Harmony. Specimens in the State Normal School collection appear to me to belong here, and these were taken at Terre Haute and at Denver, in Miami county. Others are in Mr. Smith's collection made in Fountain county. Professor Blatchley reports both *saurita* and *faireyi* at Terre Haute. Mr. W. O. Wallace, of Wabash, sends me a specimen having 21 rows of scales. The lateral stripes are on the fourth and fifth rows of scales, instead of the third and fourth, as usual. Ventrals 170; subcaudals 110.

I am not aware that the variety *proxima* has been captured in Indiana, although some have been reported. These different forms intergrade so completely that to me it appears to be useless to maintain their specific identity. It is often difficult to distinguish to which variety the specimen in hand belongs. As the species is traced from its eastern limit to the south and west it becomes stouter, the ventrals increase somewhat, and the colors deepen.

Habits.—*Eutainia saurita* has the reputation of being aquatic in its habits. This, however, can not mean that they affect the water so exclusively as do the species of *Natrix*. In fact, it leaves the water and travels considerable distances in search of food. I have taken this species frequently in the South, and I do not remember of having ever seen it in the water. However, it is found in the vicinity of streams, and no doubt often enters them. It is an active animal, as is shown by its slender and lithe form. It is wholly harmless. DeKay states that it sometimes climbs trees. He quotes Dr. J. P. Kirtland as saying that it seeks the most retired woods for its residence. Its food consists of insects, frogs, toads, and similar small animals. Of its breeding habits not much is known. In a specimen taken in Mississippi I find nine eggs with the development just begun. Professor F. W. Putnam (*22*, ii, 134) says that a female that he examined on July 13 contained nine eggs, each three-fourths inch long. In these he found embryos 2.5 inches long. On July 30 another was caught which had excluded a part of her brood, there being only four eggs left, and these were ready to be burst by the young. These eggs were 1.25 inches long, and contained young 5.5 inches. Professor G. B. Goode (*84*, '73, 182) received the following from Herman Strecker, of Reading, Pa.: "Some years ago I came across a garter snake (*Eutainia saurita*) with some young ones near her. Soon as she perceived me she hissed and the young ones jumped down her throat, and she glided beneath a stone heap. Another time I caught a snake of the same species, but, as I thought, of immense size, which I took home and put in a cage. On going to look at her, some short time afterward, I discovered a great number of young ones (about thirty, if I recollect rightly), and whilst I was still looking at the sudden increase two more crept out of the old one's mouth, and finally, after a little while, a third one did likewise." Within the body of a specimen of *fairyei*, 28 inches long, sent me from Fountain County by Mr. C. H. Smith, I find twelve eggs. Of these the hinder four are in the left oviduct. The eggs show no evidences that development had begun. The length of each is about .75 inch, the short diameter about .32 inch. The form is irregular on account of mutual pressure. In a specimen taken at Vicksburg, Miss., about July 4, I find twenty fully developed young snakes. Each is about 9 inches long.

Eutainia radix, B. & G.

Racine Garter Snake.

Eutainia radix, Baird and Girard, 1853, 6, 34; *Eutænia radix*, Coues and Yarrow, 1878, 9, iv., 277; *Eutænia radix melanotænia*, Cope, 1888, *3*, xi, 400.

A snake of moderately stout form, the tail forming from a fifth to a fourth the total length. Head distinctly marked off from the neck. Upper labials seven (rarely eight), the eye over the third and fourth. Lower labials ten, the sixth largest. Anteorbitals one, postorbitals three. The scales are in either nineteen or twenty-one rows. They are prominently keeled, so that the snake is distinctly rough in appearance. The ventral plates vary from 153 to 162; the subcaudals from 51 to 80.

Color above light or dark olive-brown, with three stripes of yellow and series of black spots. The dorsal stripe occupies the median row of scales and the adjacent half of the next row on each side. The lateral stripe lies on the third and fourth rows of scales on most of the body, but posteriorly narrows and descends to lower rows. One or two series of spots are usually to be seen below the lateral stripe, on the first and second rows. Between the lateral stripe and the dorsal may occur two series of black spots. They may be distinct or obscure The upper surface of the head is dark olive, with the usual two minute yellow dots on the occiput. The upper labials are yellow, with an edging of black. Lower jaw and throat yellow. Abdomen greenish or olive. On the outer ends of the ventrals there is, on each side, a row of black spots, and the posterior edge may be wholly edged with the same color.

Eutainia radix subspecies *melanotaenia* has recently been described by Prof. E. D. Cope from a specimen furnished him by Mr A. W. Butler, and taken in southeastern Indiana. In this variety there are 21 rows of scales. The dorsal band is nearly completely bordered on the sides with black. The space between the dorsal and the lateral stripe is nearly covered by the two rows of distinct quadrate black spots. The surface between the spots themselves is dark olive-brown. Below the lateral stripe are two rows of alternating black spots, one on each row of scales. These sometimes coincide and unite. Each of the ventrals has on each of its outer ends a black spot, and these are sometimes so large that they unite to form an interrupted stripe along the flank. Between these spots the posterior edges of the plates are black. The specimen of *Eutainia radix* reported by myself in 1881 as occurring in Indiana (*22*, xv, 738), and included in my list (*94*, '87, No. 44), proves on further examination to belong to Prof. Cope's *melanotaenia*. It differs in having 160 instead of 153 ventral plates, and 80 instead of 68 subcaudals. There is but a single row of spots below the lateral stripe, but they are large, quadrate, and lie on both the rows of scales. It was taken at Irvington.

Habits —Not much is recorded concerning the habits of this species. Drs. Coues and Yarrow found it in considerable numbers in Dakota and Montana. The were observed to stay principally about the borders of streams and pools of water, where they could catch tadpoles, young frogs, and various water insects. They are themselves preyed on by hawks. They are less active than some of the slenderer species, and when handled make little resistance. Only the larger individuals make any attempt to bite. The males and females were found pairing in September and part of October, while the greater part of the females were found to be pregnant in July and August, producing thirty or forty young. This indicates a period of gestation of nearly a year.

Eutainia sirtalis, (Linn.).

Garter Snake; Striped Snake.

This is a snake which has a very wide distribution, and of which many varieties are recognized. Many of these have been by various authors regarded as distinct species.

Eutainia sirtalis varies in form from slender to quite stout. The head is distinctly marked off from the neck. The tail makes up from a fourth to a fifth the entire length. The head is broad behind, narrowed in front. Crown-shields 9, all well developed. Anteorbital 1. Postorbitals 3 or 4. Labials 7 or 8, the eye over third and fourth. Lower labials 10. The scales are keeled, and arranged in 19 rows. The ventral plates vary in number from 140 to 180. The subcaudals from 50 to 90 pairs. The anal plate is entire.

The ground color varies from a light olive to black. This is usually relieved by three greenish or yellow stripes, a dorsal and two lateral, and by series of dark spots, some above, some below, the lateral stripes. But of the stripes and spots some or all may be missing. The belly is greenish or slate-color, usually with a series of black dots at the outer ends of the ventral plates.

Variety *sirtalis*.

Coluber sirtalis, Linnæus, 1758, *64*, ed. x, 222; *Eutainia sirtalis*, Baird and Girard, 1853, *6*, 30; *Tropidonotus sirtalis*, Holbrook, 1842, *54*, iv, 41, pl. 11; Garman, 1883, *13*, 24, pl. 3, fig. 3.

Dorsal stripe narrow, being encroached on by dorsal spots; lateral stripes not conspicuous; color above the faint lateral lines brown to black, this containing three series of indistinct spots, each of about 70 from head to vent; color immediately below lateral lines greenish; abdomen with a row of black blotches on the ends of the ventral plates on each side. All North America that is inhabitable by snakes.

Variety *dorsalis*.

Eutainia dorsalis, Baird and Girard, 1853, *6*, 30; *Eutænia sirtalis*, subsp. *dorsalis*, Cope, 1875, *12*, 41.

All three stripes broad and conspicuous; space above lateral bands bright olivaceous, and containing a series of rather large spots, 74 in number, from head to vent; abdomen greenish, with a black dot on each end of each ventral plate. Maine to Florida and Mexico.

Variety *ordinata*.

Coluber ordinatus, Linnæus, 1766, *64*, ed. xii, i, 379; *Eutainia ordinata*, Baird and Girard, 1853, *6*, 32; *Eutænia sirtalis* subsp. *ordinata*, Cope, 1875, *12*, 41; *Tropidonotus ordinatus*, Holbrook, 1842, *54* iv, 45, pl. 12.

Stripes all indistinct; space above the lateral bauds olive, with three distinct rows of square dark spots, about 85 in number, from head to vent; abdomen like the above varieties.

Variety *parietalis*.

Coluber parietalis, Say, 1824, *14*, i, 186; *Eutainia parietalis*, Baird and Girard, 1853, 6, 28; *Eutænia sirtalis* subsp. *parietalis*, Cope, 1875, *12*, 41; Coues and Yarrow, 1878, *9*, iv, 276.

Three distinct stripes, a dorsal pure yellow; the lateral a heavier yellow, tinged more or less with red; color between the stripes deep brown to black; spots, if any, indistinct; skin between the scales brick or vermillion red, this tingeing many of the scales. The red is most distinct just above the lateral stripe, but also extending down to the ventral plates. Belly whitish green, with black spots on the ends of the ventral plates. Length 3 feet. Indiana to California.

A specimen of *E. sirtalis* sent me by Mr. Chas. Beachler, of Crawfordsville, agrees well with the description given by Prof. Baird, and more especially with that given by Coues and Yarrow, as cited above, except that the red, which is so characteristic of this variety, is distinctly seen only on the anterior third of the body. Spots on ends of ventrals rather large. Specimens are frequently taken about Irvington which are referred to this variety. Prof. Blatchley states that he has taken *parietalis* at Terre Haute. Mr. A. B. Ulrey has shown me a specimen from North Manchester, Wabash county, which conforms to the description of *parietalis*, except that it has twenty-one rows of scales. The red in Indiana specimens does not extend to the posterior portion of the body, and it is more abundant and extends further back on specimens from the northern part of the State than on those taken about Indianapolis.

Variety *obscura*.

Eutainia sirtalis obscura, Cope, 1875, *12*, 41; Cope. 1888, *3*, xi, 399.

This is without spots, uniform brown between the bands. It is said (*3*, 11, 399) to resemble *E. saurita* at first sight. A specimen shown me by Mr. A. B. Ulrey, from North Manchester, Wabash county, appears to be referable to this variety.

Eutainia sirtalis graminea, Cope, 1888, *3*, xi, 399.

This interesting variety has been described by Prof. Cope from a specimen furnished him by Mr. A. W. Butler. It was taken in eastern Indiana. It is described as of a uniform light green above, yellow, clouded with green, below. There were no stripes nor spots on the body, nor any markings on the head. It resembled closely the Rough Green-snake, but it is to be distinguished from it by having 19 instead of 17 rows of scales, and by the entire anal plate, as well as by other generic characters.

Habits.—*Eutainia sirtalis* is probably the most familiar and best known of all our snakes. It is found everywhere over the country, and every country school-boy has possibly carried specimens of it in his pocket to the terror of his girl playmates, and has almost certainly killed the snakes by the dozen. This species is not so active as some others, but they are not usually so sluggish as some authors describe them. Holbrook says that it is exceedingly gentle and can be handled without fear of injury; indeed, he states that he has never known them attempt to bite. DeKay, on the other hand, says that they will bite and sometimes leave troublesome, but not dangerous, wounds. Both statements seem to exaggerate the usual disposition of the snakes. It is certain that they will at times strike and bite, but it must be rarely, indeed, that they produce anything worse than a few slight scratches.

These snakes are frequently found in moist situations, either near the water or even swimming about in it. They can not, however, be called aquatic. They are likely to be found in or near the water because there they can find abundant prey. They are often seen far away from the water in dry fields. Drs. Coues and Yarrow say of *parietalis*, that it is common in the clear, cold waters of lakes and streams of Dakota and Montana, and was often seen swimming freely in deep water at some distance from shore. This snake lives on frogs, toads, insects and small mammals. All observing persons have probably seen in this snake illustrations of the great size of prey that a serpent can swallow. The frog is often of a considerably greater diameter than the snake itself. One writer (*22*, ii, 50) relates an instance of a Garter-snake chasing and catching a mouse. Immediately the snake enveloped the mouse in its coils, crushed it boa-constrictor-like, and then swallowed it. Another writer (*22*, '84, 88) tells of seeing the Garter-snake in a pool catching small fishes. Shortly afterward in passing the pool it was dried up, and this snake was eating the dead fishes. On the other hand, the snake is preyed on by hawks, owls, hogs, ducks, and turkeys. I have seen a cat eat a Garter-snake.

These snakes, like others of the genus, are ovoviviparous. The number of young produced at a time is considerable. Prof. F. W. Putnam (*22*, ii, 134) informs us that on July 1, in a female 35 inches long, he found 42 nearly developed young, each of which was 5.5 inches long. Dr. J. Schneck, of Mt. Carmel, Ill., says (*22*, xvi, 1008) that from a female 35 inches long he took 78 young from 3 to 7 inches in length. They were pressed from the vent. The first twenty were free, the others confined within the egg-coverings. In a female 33 inches long, taken by Dr. Alex. Jameson near Indianapolis, about August 1, I find 39 young snakes not yet ready to be born. The food-yolk is not all used up. The egg-tooth is present. A female from Paris, Ill., contained 35 young, each 7 inches long. The food-yolk was all gone and the egg-tooth shed. It appears that the young are born late in the summer or early in the fall. The sexes are said to unite in September or in October, but it seems probable that this also occurs early in the spring. This is one of the snakes which has the reputation of swallowing the young when they are in peril Col. Nicholas Pike, who is an accurate observer, assured Dr. Goode (*34*, 1873, 182) that he had seen the Garter-snake afford its young family temporary protection in its throat, from which they were soon noticed to emerge.

On the approach of cold weather these snakes seek some opening in the earth and then become dormant In some instances they appear to collect in considerable numbers where they pass their winter slumber. We thus occasionally hear of bundles of snakes being plowed up. E. L. Ellicott relates (*22*, xiv, 206) having seen very early in the spring, in Maryland, a bundle of Garter-snakes, in which some hundreds of them could be counted. It is altogether probable that such assemblages are determined partly by the sexual impulses. The Garter-snake leaves its place of hibernation apparently as soon as the first warm days come, although they may relapse again into the dormant condition. At Irvington I have taken them as early as the 7th of March.

Like all other snakes this species at intervals sheds the outer coat of the skin. How often this occurs I do not know. Dr. Benj. Sharp (*1*, 1890, 149) observed the process of exuviation in the Garter-snake. Two specimens were kept in an aquarium. When observed one had just crawled out of the water and then shrugged and shook itself. Finding a narrow place it pressed itself in so that the skin parted along the jaws. This was pushed back behind the head. Then the snake crawled through the opening, escaping from the skin and leaving it turned inside out. The operation required less than a minute. One of the skins was without rent or loss of a scale. This occurred on April 13. Some specimens that I kept for awhile shed the skin about June 1. In case the snake can not have access to water the shedding of the skin is a more prolonged operation.

Eutainia butlerii, (Cope).

Eutænia butlerii, Cope, 1888, *3*, xi, 399.

This is described as having the head little distinct, muzzle conical, a little protuberant; eye not large. There are 144 ventral plates; subcaudals 62. The ground color is olive-brown above, and there are the usual three stripes. The median has the usual width. The lateral stripes are unusually wide, covering the second, third and fourth rows of scales. They are all bordered by black. At the upper border of the lateral band there is some indication of spots. No markings on the head and labials. The abdomen is olive, with an elongated black spot at each end of each ventral plate.

The length of the only specimen known is not given in the description. Locality, Richmond, Ind.

Family ELAPIDÆ.

Maxillary provided with an erect, immovable poison fang, which is grooved in front. Smaller teeth behind the fang or not. Head furnished with plates. Loral usually absent. Tail short and conical.

Of this family there is but a single genus known to inhabit America, namely, the

Genus ELAPS, Schneider.

Elaps, Schneider, 1801, *4 l*, 289; Baird and Girard, 1853, *6*, 21; Garman, 1883, *1.*, 104

Body elongated and cylindrical. Head little, if any, wider than the neck; its upper surface with the nine plates usually found in the *Colubridæ* No loral. Nasals two, with the nostril between, or mostly in the anterior. Anteorbital single, occasionally fused with the postfrontal. Eye small, the pupil round. Scales without keel, smooth and glossy; arranged in thirteen to fifteen rows. Anal plate divided, rarely entire. Subcaudals divided

A genus with numerous species. Its members range from the Argentine Republic to the United States, where there are two or three species. One of these, the following, is found from South Carolina to Texas and Mexico. Recently it has been discovered in Southeastern Indiana.

Elaps fulvius, (Linn.).

Coral Snake.

Coluber fulvius, Linnæus, 1766, *64*, 381; *Elaps fulvius*, Cuvier, 1817, *124*, ii, 84; Holbrook, 1842, *54*, iii, 49, pl. 10; Baird and Girard, 1853, *6*, 21; True, 1883, *22*, xvii, 26.

Body elongated, slender and cylindrical. Head little wider than the neck, flat above. Loral absent. Anteorbital 1; postorbitals 2. Upper labials 7, the eye over the third and fourth. Lower labials 7, the fourth large. Scales smooth, arranged in fifteen transverse rows. Ventral plates 202 to 236. Subcaudals 25 to 44 pairs.

The colors are black, bright red, and yellow. These are arranged in bands which encircle the body. The head, from the hinder end of the vertical plate to the snout and to the tip of the lower jaw, is black. The remainder of the head is encircled by a band of bright yellow. Behind this there follow, alternately, bands of black and red, about 13 to 20 of each, to the vent. Each red band is separated at each end from the contiguous black band by a narrow ring of bright yellow. The tail is alternately black and yellow, there being about four rings of each. In the red bands surrounding the body is usually found a number of scales of a brown color. There may also be one or more small blotches of black in the middle of the ventral portion of the red bands.

The Coral-snake is doubtless the most highly colored and beautiful snake that is to be found in the United States. It is moderately common in extreme Southern States, but, to our surprise, it has recently been taken in southeastern Indiana. Prof. A. J. Bigney, of Moore's Hill College, has shown me a specimen that was captured about a year ago two miles from the village of Milan, in Ripley county. The specimen is a typical *Elaps fulvius*, 22.5 inches long. It has 212 ventral plates and 45 subcaudals. It appears wholly unlikely that this specimen was accidentally introduced where it was found. That the species is really an inhabitant of that region is rendered still more probable from the fact that another specimen was taken near there, in Hamilton county, Ohio. It is preserved in the collection of the Cincinnati Society of Natural History (A. W. Butler in *94* for 1892). This occurrence of the Coral-snake in that portion of Indiana indicates that it will be found all along the southern border of the State, and thence down the Mississippi River to the Gulf.

Habits.—The Coral-snake appears to be somewhat sluggish in its disposition. It is poisonous, but it is said to bite only when provoked. In the American Naturalist for 1883, page 26, Prof. True has given us evidence of the ability of the snake to inflict dangerous bites. Several cases are reported, some of which proved fatal. However, on account of the smallness of the serpent's mouth, the shortness of its poison-fangs, and their backward position, it seems probable that dangerous wounds could be inflicted only on some such organ as the fingers or the toes. I do not know anything about the breeding habits of this genus. The Coral-snake appears to have the habit of capturing and eating other serpents. In volume xxii of the *Encyclopedia Britannica* is an illustration of one which is swallowing another snake almost as large as itself. In the alimentary canal of a specimen 21 inches long, from Florida, I have found a *Storeria* 13.5 inches long.

Persons who are living in the southern portion of the State ought to be on the watch for this strikingly colored snake, and all specimens of it should be sent where they will be preserved for study.

Family II. CROTALIDÆ.

Form usually thick and short. Head triangular, broad behind, flat, and distinctly separated from the body by a small neck. Maxillary bone much shortened, moving freely on the lachrymal and supporting a single functional, enlarged, tubular tooth, or poison fang, which is capable of erection and of concealment under a fold of the lining of the mouth. A deep pit between the nostril and the eye. Poison glands at the sides of the head behind. Pupil oblong, perpendicular. Scales keeled. Anal plates entire. Tail short. All venomous.

This family contains both Old and New World genera, but most of them belong to North and South America. They are of great interest and have been much studied on account of their many peculiar structures, and because of the deadly effects of their venom. Those who are interested in learning more about the anatomy of these dreadful serpents may consult such works as Huxley's Anatomy of Vertebrates, the various encyclopedias, and volumes XII and XXVI of the Smithsonian Contributions to Knowledge. An interesting popular summary of the results of the studies of Drs. Mitchell and Reichert may be found in the Century Magazine for August, 1889 (vol. XVI, 503).

KEY TO THE U. S. GENERA OF *Crotalidæ*.

A. Tail not provided with a rattle. *Agkistrodon*, p. 123.
AA. Tail provided with a rattle.
    a. Upper surface of head with the usual 9 crown-shields.
        Sistrurus, p. 125.
    aa. Vertical and occipital plates replaced by small scales like those of the back. *Crotalus*, p. 128.

Genus AGKISTRODON, Beauv.

Agkistrodon, Pal. de Beauvais, 1799, *36*, iv, 381; Baird and Girard, 1853, *6*, 17; *Ancistrodon*, of various authors.

Head flat, triangular, with 9 crown-shields. Loral present or absent. Scales in 23 to 25 rows, all keeled. Anal plate entire. Subcaudals not divided except some of the posterior. No rattle.

Loral present. Scales in 23 rows. *contortrix*, p. 123.
No loral. Scales in 25 rows. *piscivorus*, Appendix.

Agkistrodon contortrix, (Linn.).

Copperhead.

Boa contortrix, Linnæus, 1766, *64*, ed. xii, 373; *Trigonocephalus contortrix*, Holbrook, 1842, *54*, iii. 39, pl. 8; *Agkistrodon contortrix*, Baird and Girard, 1853, *6*, 17; *Ancistrodon contortrix*, Garman, 1883, *13*, 120, pl. 8, fig. 1.

Head large, flat, triangular, and with the sides in front of the eyes perpendicular. Neck slender. Tail short, about one-eighth the total length, ending in a curved horn. Vertical plate as broad as long, pentagonal. Occipitals about the size of the supercillaries, showing a tendency to break up into smaller scales. Both pairs of frontals well developed, extending out to the edge of the upper surface of the head. A small plate sometimes lying between the vertical and the postfrontals. Two anteorbitals, the lower smaller and bounding the pit above. A loral present. Nasal divided, with the nostril present. Rostral broad and high. Upper labials 7 or 8, excluded from the orbit by a number of suborbitals, the second with a groove leading into the pit. Inferior labials 9-10. Scales in 23 rows, strongly keeled. Ventral plates about 150. Subcaudals 42 to 52, all entire except the last 8 to 18.

Spotted, with black, brown, red, reddish yellow and yellowish white. On the sides and reaching up to the middle of the back are blotches 6 scales long of hazel brown, margined with dull red. These blotches are narrow below, but run into the colors of the belly. Between these spots are larger ones of smoky brown, sometimes with a good deal of chocolate, and with dark margins and lighter centers. These widen below, but become narrower toward the middle line. On the middle of the back they either alternate with those of the other side or become confluent with them, thus forming bands across the body. The end of the tail is black above. Lower surface yellowish, more or less clouded with brown and with large black spots at the ends of the ventrals. These cover the ends of two ventrals and are separated by the ends of three. Head above reddish or copper-colored, this color extending down on the sides to the level of the corners of the mouth and lower borders of the eyes. Below this the color is cream yellow. Lower jaw paler, but with reddish brown streak along the lower edge of the inferior labials. Length probably sometimes three feet. Terrestial.

Habitat. Massachusetts to Florida, west to Kansas and Texas. In Indiana this venomous serpent, once abundant in most localities of the southern half of the State, is now happily becoming rare; in most localities it is probably entirely exterminated. Where, however, the country is not thickly settled, and where there are abundant forests and rocks, it may even yet be found in considerable numbers. About New Harmony the collectors have frequently found it. It has been taken in Monroe County (Bollman). Mr. C. H. Smith reports its occasional appearance in Fountain county, and Mr. Hughes tells me that one was found, the first in many years, in 1889 in Franklin county. In the northern portion of the State it has probably always been scarcer, but still present.

The habits of this snake are tolerably well known. Unlike its equally disreputable relative of the South, it is terrestrial in its inclinations. However, its desire for water to drink and the better opportunity to obtain prey will often lead it to the neighborhood of ponds and streams. In its usual movements it is a sluggish animal, depending more for safety on the terror that it inspires and on its fangs than on flight. It appears to prefer to lurk about dark and shady situations. Its food consists of small birds, mammals, and frogs. It is itself devoured at times by other snakes, as the black-racer and the water-moccasin (*22*, iii, 158; *44*, '77, 399). In securing its prey it depends on the deadly effects of its poison. Mr. S. Garman states that they usually eat the prey as soon as it is dead, and even before it ceases to struggle. Sometimes lively mice would elude two or three strokes, and this would seem to throw the snake into an ecstacy of excitement. They would not eat fishes.

These snakes are greatly dreaded on account of their biting without giving any warning. However, some warning they do sometimes attempt to furnish. This is done by rattling on objects by quick vibrations of the tail. Many species of snakes are now known to do this. Whether their bright, dappled colors are intended to act as a warning to large animals not to tread on them, or are to protect the snakes from the observation of their prey as they lie among dead forest trees, is hard to decide. Mr. Garman says that they do not yield to good treatment. Some that he observed, though they were frequently and gently handled, remained vicious and ready to bite. These snakes are very poisonous, but the researches of Drs. Mitchell and Reichert show that their venom is less virulent than that of the rattlesnakes, but more so than the water-moccasin. While the introduction of the poison into the system produces important changes in the appearance of the skin, due to changes in the blood and blood-vessels, and though after recovery some of these appearances might recur, yet the stories that the limb becomes spotted just like the snake that did the biting, that it annually feels the bite and regains such colors, are wholly absurd.

As regards their breeding habits, the copperhead, so far as known, produces living young. The number of young produced each year is variously given. According to Professor J. A. Allen, five out of seven females caught in the latter part of July, in Massachusetts, contained slightly developed embryos, while of six females killed in September the oviducts of each contained from seven to nine young six inches long. The sexes, sometimes at least, pair in August. Other writers, principally, however, through the newspapers, put the number of young as high as 60 to 80. This is undoubtedly an error, which probably arises from confounding the copperhead with other snakes, especially with the hog-nosed snake, *Heterodon platirhinos*. This is another of the species reported to Dr. Goode (*34*, '73, 184) as furnishing to its young in times of danger a retreat down its throat.

Genus SISTRURUS, Garman.

Crotalophorus, Gray, 1825, *114*, 205; *Sistrurus*, Garman, 1883, *13*, 176.

Form short and stout. Tail short. Head with nine crown-shields, as in the *Colubridæ*. Loral present. Tail furnished with a rattle.

Sistrurus catenatus, (Raf.).

Massasauga; Prairie Rattlesnake.

Crotalinus catenatus, Rafinesque, 1818, *80*, iv, 41; *Crotalus tergeminus*, Say, 1823, *14*, i, 499; *Crotalophorus tergeminus*, Holbrook, 1842, *54*, iii, 29, pl. 5; *Caudisona tergemina*, Wagler, 1830, *75*, 176; *Sistrurus catenatus*, Garman, 1883, *13*, 176, pl. 9, fig. 2; *Caudisona tergemina*, Cope, 1875, *12*, 34; *Crotalophorus catenatus*, Cope, 1892, *3*, xiv, 684.

Body stout. Head short. Tail constituting from an eighth to one-eleventh the entire length. Head covered above with the nine crown-shields usually found in the non-poisonous snakes. Anteorbitals 2, the upper horizontally elongated, joining the prefrontal and excluding therefrom the small triangular loral. Lower anteorbital narrow and long, forming a part of the boundary of the pit. Supraorbitals large and overhanging the eyes. Upper labials 11 to 14, excluded from the boundary of the orbit. Lower labials 12, small. Scales in 25, rarely 23, rows, all strongly keeled, except those of the outer row. Ventral plates 136 to 153. Subcaudals 21 to 29, a few of the most posterior divided.

The ground color varies from light ashy, often with a reddish tinge, through ashy brown to deep brown and even black; below from yellowish, sprinkled with dusky, to slate blue, mingled with yellowish white and black. There are above, seven series of dark spots, which are variable in intensity of color from chestnut-brown to black. When the ground color is very dark the spots may disappear, and we see then a uniformly black snake. The outer borders of the spots are usually darker than the centers, and are surrounded with a narrow line of paler. The spots of the successive rows alternate. Those of the median, or dorsal, series are from 35 to 48 in number, and descend on each side to about the eighth row of scales. They are broader than long, and in the median line are notched in front and behind. Occasionally they may be wholly divided. Just below this series is another of much smaller spots. Further out is a series of spots which extend from the second to the sixth row of scales. The lowest series on each side lies on the outer ends of the ventral plates and the three outer rows of scales. Each side of the head with a broad, dark band from the eye back on the neck. This is bounded below by a streak of yellowish white from the pit to the corner of the mouth and on the neck. Upper labials mostly dusky. Upper surface of the head with two dark, longitudinal streaks and three paler ones. Lower jaw variegated with yellowish white and dusky. A pale transverse band from eye to eye, bordered with darker. The length may reach three feet in fully grown specimens.

Ohio and Michigan to Utah. Said by Garman to extend to Mississippi, on what authority I do not know. In Indiana it is abundant in some localities, but I have not been able to confirm its occurrence south of Indianapolis. Wabash county (D. C. Ridgley). I have seen specimens from Laporte, Hendricks, Hamilton and Montgomery counties. They appear to be abundant in the swampy grounds in the neighborhood of Lake Maxinkuckee, in Marshall county. The black specimens are frequently found in Indiana. They were once described as a distinct species, but their dark coloration is probably nothing more than an individual variation. We have a very similar case in the differently colored forms of *Heterodon platirhinos*, *Coluber obsoletus*, and *Natrix sipedon*.

HABITS.—This species is, on an average, considerably smaller than the Banded Rattlesnake, *Crotalus horridus*. It is, on that account, less to be feared than that serpent, since the fangs would naturally penetrate less deeply, and the amount of poison that is injected into the wound would be less. Indeed, Dr. Kirtland, of Ohio, is quoted as saying that its bite is scarcely worse than the sting of a hornet. But having had a good deal of experience with, and knowledge of, these snakes, I think that they are not to be tampered with. Animals that have been bitten by them, such as dogs and cows, suffer much, and have troublesome swellings. The rattle is less powerful than that of its larger relative, but may be heard at a sufficient distance. The snakes appear to prefer low, wet grounds as their habitation, but they are not aquatic. Yet they may often be found far away from water in dry fields. On the prairies of Illinois, before the country became thickly populated, these reptiles were extremely abundant, and the killing of two or three dozen of them in a season was not an unusual thing for any farmer's boy. Now, in that same region, not one is seen in years. This disappearance of these snakes has been supposed to be due to the destruction wrought among them by hogs. Yet, on those prairies, in those days, there were no roaming hogs. The extinction of the snakes may be due to the breaking up of the soil, the draining of the ponds, and the clearing away of the rank vegetation, which furnishes them protection. At the present day it is only in swamps and marshes that they are found.

It appears that these snakes shed their skins at least twice a year; and since, further, Garman has shown that the segments of the rattle represent a retained portion of the sloughed epidermal covering, it seems quite probable that two or more joints of the rattle are produced each year. In any case, the age can not be determined by the number of segments, since the terminal ones are continually being worn off and lost.

The young of this species are brought into the world alive. They are about six in number at each brood, and when born are about six inches long. They appear about the first of September. This species has been included by Goode in his list of those whose females allow the young a place of safety in the stomach. The writer has published an account of the observations made on two females of this snake by a man of credibility who had captured them and kept them until they had produced young (22, xxi, 216). According to these observations, the young passed freely into and out of the mother's mouth until they were a month old. After this time the mother was very attentive to the young, as I saw myself.

Genus CROTALUS, Linn.

Crotalus, Linnæus, 1758, *64*, ed. x, i, 214; Baird and Girard, 1853, *6*, 1; Garman, 1883, *13*, 110.

Form stout. Head triangular. Crown-shields fewer than nine, the vertical and occipitals replaced by numerous small scales. A pit in front of nostril. Anal and most of the subcaudals entire. Tail with a rattle.

Crotalus horridus, Linn.

Banded Rattlesnake.

Crotalus horridus, Linnæus, 1758, *64*, ed. x, i, 214; Garman, 1883, *13*, 115, pl. 9, fig. 1; *Crotalus durissus*, Holbrook, 1842, *54*, iii, 9, pl. 1; Baird and Girard, 1853, *6*, 1.

Body elongated, tapering toward head and tail. The neck slender. Tail short, from one-eighteenth to one-eighth the total. Head broad, triangular, flattened above, concave in the interorbital region. Snout blunt. Back portion of head forward to the interorbital region covered with small keeled or tuberculated scales. Scales of the cheeks larger and smooth. Superciliaries large and overhanging the eyes. Anteorbitals 2, the lower slender and overarching the pit. Lorals 2. Nasals 2, the anterior largest. Rostral higher than broad. Prefrontals 2, postfrontals 4, in a transverse row, the inner pair sometimes replaced by small scales. Upper labials 12 to 16, separated from the orbit by three or four rows of scales. Lower labials 13 to 18. Eye small. Scales keeled, except those of the outer row; number of rows 23 or 25. Ventral plates 165 to 175. Subcaudals 19 to 25. Rattle of the adult parallelogrammic, with a median groove.

Ground color above cream yellow to yellowish brown, and even black. There are three series of dark spots, a dorsal and a lateral on each side. The dorsal blotches are large, occupying in breadth about twelve rows of scales and about four in length. These are nearly black around the borders and paler in the centers. The lateral series has the spots smaller. Posteriorly the three series coalesce, so as to form zig-zag, dark-edged, transverse bands. All the spots and bands are bordered with sulphur yellow. Tail of the adults usually black. The color below is yellow, with some mottlings and sprinklings of black. Upper lip sulphur yellow; lower lip paler.

The length of this species may become as much as six feet, possibly more.

Distributed from Massachusetts to Kansas, south to North Carolina and Texas.

It is to be found, in all probability, in nearly all the counties of Indiana, but it is in most places quite rare. I can name only two localities where it has been taken recently and record of it preserved. These are New Harmony (Sampson's coll.) and Monroe county (Ind. Univ. coll.).

HABITS.—No American snake has probably been so carefully studied as has this and its immediate relatives. This is due to the dangerous nature of this animal and to the interesting structures shown in its rattle and its arrangements for producing and instilling its poison. The results of some of these studies may be learned by consulting the works referred to on page 123.

In its free state this species appears to inhabit wooded districts, although it may probably sometimes be found on prairies. It especially delights in taking up its abode where there are rocks and debris among which it can find at short notice a safe retreat. Its movements of locomotion are rather slow. When surprised it will often seek to escape, without inflicting injury on its enemy. When, however, it is pressed, or there is no time for retreat, it delivers a blow with such rapidity that the motion can hardly be followed. The mouth is held open, the fangs directed forward, and if possible they are buried in the victim. At the same time the poison gland is squeezed by the proper muscles, so that the poison is injected deep into the wound. If the amount of poison is large it may be quickly fatal to even large animals. Small animals, as birds and mice, almost immediately succumb to the deadly influence. It is usual for the Rattlesnake to sound its rattle when it has been disturbed by some animal which it has reason to fear. The use of this alarm has been much discussed. Some have regarded it as an imitation of grasshoppers in order to allure birds within its reach. Others have thought it a sexual call. Still others think it a providential arrangement to prevent injury to innocent animals and man. It is doubtless of use to warn off animals that might do injury to the snake itself, or at least compel it to use up its store of poison and its fangs, all of which it needs to procure its food. Dr. A. R. Wallace has suggested in his recently published "Darwinism" that the creature has acquired the structure and habit in order to warn off buzzards and other snake-eaters that might pounce on it as it lies on naked rocks. It is a warning note, saying, "Look out for yourself! Your life if you injure me."

Rattlesnakes do not appear to try to injure one another by biting. Indeed, Dr. Mitchell states that the poison does not affect the snakes hemselves. He says that he has over and over injected under the skin of a rattlesnake its own venom or that of a moccasin or of another rattlesnake, but he had in no case seen a death. He often kept from 10 to 35 rattlesnakes together without any of them harming the others. If a large snake were suddenly dropped on the others they would show no resentment, whereas if any other animal were thus dropped on them it would immediately get a blow. In captivity they are extremely sluggish, not moving, and refusing for long periods to accept food. Usually, after about a year without food, they will kill and eat animals. Dr. Mitchell fed his numerous specimens by putting a long funnel down their throats and pushing the food into their stomachs. They were very fond of water, and would drink large quantities of it and lie in it for hours.

They shed their skins at different times. If they did not have water the skin would come off in patches. He says nothing about the relation of the shedding of the skin and the acquisition of new segments of the rattle. It has been noted by observers that a variable number of segments of the rattle is acquired each year, although the popular idea is that one is the number. As high as four have been observed to be added in a single year. The terminal segments, too, are constantly being worn off, so that the number of segments present is no indication of the age of the snake.

On each side of the upper jaw is to be found a single fang, which is solidly attached to the maxillary. These are probably shed at intervals, and besides are liable to be torn out in use. Alongside of the functional fang there may be found a number of reserve fangs, as high as seven sometimes, the oldest of which moves into the place of the lost fang and soon becomes ankylosed to the bone. Holbrook states that Mr. Peale, of the Philadelphia Museum, kept a living female rattlesnake for fourteen years. She had eleven rattles when she came into his possession. Several were lost annually, and new ones formed. When she died there were still eleven. During this period the snake had grown four inches in length.

This snake, like all the *Crotalidæ*, brings forth its young alive. The number appears to be about nine. I found this number of eggs in a female 37 inches long, brought from Pennsylvania. The eggs were 1.5 inches long by an inch in short diameter. Of these there were four in the left oviduct. There were evidences that development had begun.

Where these snakes are numerous they are inclined to gather in considerable numbers in caverns in rocks and similar places, in order to undergo their winter sleep. Such places form the rattlesnake dens about which we hear occasionally.

Order LACERTILIA.

Lizards.

Animals varying in form from stout and almost toad-like to long, slender, and snake-like. Limbs usually present, but absent in a few forms, as in our Glass-snake. Vertebræ few to many. Bones of the brain-case not so firmly united as in the snakes; those of upper and lower jaws less loosely connected than in the serpents, as a result of which the mouth is not distensible. Mandibles united at symphysis by suture. Teeth present. Skin usually provided with granular or overlapping scales. Vent transverse slit.

A large and widely distributed order, divided by Dr. Boulenger into wenty-one families. Of these four are represented by species in Indiana.

KEY TO THE FAMILIES OF INDIANA *Lacertilia.*

A. Whole tongue or its posterior larger portion covered with close-set papillæ, like the pile of velvet.
  a. Tongue thick, little free, not or but feebly notched in front, wholly covered with villiform papillæ. Limbs present. No bony plates underlying the scales. *Iguanidæ*, p. 131.
  aa. Tongue posteriorly thick and with villiform papillæ; anteriorly thin, free, deeply notched, and covered with scale-like papillæ. Scales underlaid with bony plates.
  Anguidæ, p. 134.
AA. Tongue covered with scale-like papillæ.
  a. Tongue free in front, ending in two long points. No bony plates underlying the scales. *Teiidæ*, p. 136.
  aa. Tongue rather long, free in front and sides; slightly notched in front. Bony plates underlying the scales.
  Scincidæ, p. 138.

Family III. IGUANIDÆ.

Lizards varying in form from short and thick to elongated and slender. All with functional limbs. Dentition pleurodont. Tongue thick, fixed to the floor of the mouth or slightly free in front. The tip slightly or not at all notched. Plates of the upper surface of the head usually small, but well developed in our genus *Sceloporus*. No bony plates beneath the epidermal scales.

A large family, containing, according to Boulenger, 50 genera and about 300 species, with few exceptions inhabiting the New World. Only one species known in Indiana.

Genus SCELOPORUS, Wieg.

Sceloporus, Wiegmann, 1828, 77, 369; Boulenger, 1885, 29, ii, 216. Body depressed; tail long. Head with the plates of the upper surface well developed; the most posterior and median one (occipital) largest. No bony spines on head. Anterior teeth simple, the lateral tricuspid. Tympanum distinct, sunken. No gular folds. A series of pores (femoral) on the under side of the thigh. Scales overlapping and about equal in size. Contains about 25 species inhabiting North and Central America.

Sceloporus undulatus, (Bosc.).

Alligator Lizard; Pine Tree Lizard.

Agama undulata, Bosc, 1803, 69, iii, 384; *Tropidolepis undulatus*, Holbrook, 1842, 54, ii, 73, pl. 9; *Sceloporus undulatus*, Fitzinger, 1843, 82, 75; Boulenger, 1885, 29, ii, 227.

Head broad and flat, tapering to the short and rounded snout. Neck moderate, without gular fold. Tail considerably longer than the head and body, and slender. Head with the plates of the upper surface rather large. A series of these over each eye. Back of the head with three plates, the median (occipital) larger than the lateral (parietal), and with a central translucent spot. Anterior border of the ear denticulated with three or more scales.

Hinder limb pressed to the side reaching about to the ear. Distance from the base of the fifth toe to the tip of the fourth greater than from the tip of the snout to the ear. Under side of the thigh with a row of from 13 to 17 pores. Dorsal scales conspicuously keeled, the keel running out into a sharp point. Ventral scales smaller and smooth. Scales of the sides forming rows that run upward and backward. There are from 38 to 50 scales around the body and about 35 from the occiput to the base of the tail. The tail is covered with large scales, with the keel running obliquely and ending in a point.

The color above is olive or brownish, sometimes almost black. Across the back are undulating lines of dark brown or black. On each side of the back there is a row of whitish spots. The tail is cross-lined with black. Sides mottled with black, and with some whitish spots. The male has a green or dark blue spot on each side of the throat, sometimes almost black. The color is prolonged on to the arm. Each side of the abdomen is green or steel blue. The front of the thigh is dark blue or black. Middle line of the belly yellowish, with mottlings of black. The female has the throat blue, the abdomen yellowish, with black markings. Dr. C. C. Abbott says that the colors are not indicative of sex.

The length of this species becomes about 6 inches.

Distribution from New Jersey to Oregon and south to Texas. In Indiana it has been taken at Brookville (Dr. Haymond); Bloomington (Bollman); Brown county (Nor. Sch. Coll.). It is said to be rather common about Brookville, but rare in the neighborhood of Bloomington. Specimens were also seen by the author in Crawford county. One was brought from near Wyandotte Cave by W. P. Hay. Clark and Jefferson counties (Butler). I have seen a specimen said by Mr. Fletcher Noe, of Indianapolis, to have been taken by him near the city. It will probably be found to occur throughout the southern half of the State, at least in localities where there are timber and sand, and an abundance of rocks. In the South they are most abundant in the pine forests. On this account they have received the name of "Pine-tree Lizard."

These little animals are extremely active and they are able to run with great swiftness. Holbrook says that they are often found under the bark of decaying trees. It chooses also old fences as its basking places. It is given to climbing trees in search of insects and for safety from pursuers. DeKay states that when irritated in confinement they elevate their spinous scales in such a manner as to present a very formidable appearance. They are perfectly harmless, although they are often regarded as venomous. DeKay further says that they are able to alter their colors, the back assuming an azure tint.

Dr. C. C. Abbott, of New Jersey, has studied to some extent the habits of this interesting animal. They were not disposed to seek safety from pursuers in flight, but rather by concealment. They would dodge behind the trunk of a tree, and might be caught in one hand while trying to avoid the other. On being seized they do not fight and bite, but become almost instantly tame. Where they were abundant the children made pets of them and were never bitten. The lizards are extremely susceptible to changes of temperature. They are most active at about 100° F. At lower temperatures they become sluggish; at temperatures much higher they seek to escape the direct rays of the sun, and if the heat is still increased they are thrown into a state of lethargy. Their color appears to be highly protective, since they closely resemble the rough gray bark on which they so often rest. Abbott came to the conclusion that their vision is not acute, while their hearing is sharp. In endeavoring to catch flies they often missed their aim, although the insects were within easy reach. Their food consists of flies, ants, small spiders and the like. Some of Dr. Abbott's experiments tend to show that the so-called "pineal eye" is yet sensitive to the light.

The eggs are said to be laid in the sand, probably in little groups. They are deposited about June 1, and are hatched about July 10. The eggs are long and narrow, are covered with a tough coat, and are without any calcareous material. The egg weighs about 20 grains. They are abandoned to their fate, but when the young are hatched they are treated with the utmost gentleness by all the adults.

Family IV. ANGUIDÆ.

Lizards having an elongated, slender body, sometimes extremely so. Limbs present or absent; when present they may be well developed or rudimentary. Dentition various. Tongue with the posterior portion thick and covered with villiform papillæ, the anterior portion thin, flat, notched in front, and covered with overlapping scale-like papillæ. Plates of the upper surface of the head large; an unpaired occipital plate always present. Epidermal scales of the body underlaid with bony plates.

A family containing 7 genera and about 45 species. Living for the most part in the Americas, but represented likewise in Europe and Asia. We have only the following genus:

Genus OPHISAURUS, Daud.

Ophisaurus, Daudin, 1803, *69*, vii, 346; Boulenger, 1885, *29*, ii, 281; *Ophiosaurus*, Wagler, 1830, *75*, 150.

Elongated, snake-like lizards, with very rudimentary limbs or usually with none at all, the tail forming considerably more than one-half the total length. Body with a fold on each side, ending at the vent. Scales placed in straight rows longitudinally and transversely. Teeth on pterygoid, sometimes also on palatines and vomer. Teeth pointed or rounded.

This genus is represented by five species. We have only one of these.

Ophisaurus ventralis, (Linn.).

Glass-snake; Joint-snake.

Anguis ventralis, Linnæus, 1766, *64*, ed. xii, 391; *Ophisaurus ventralis*, Daudin, 1803, *69*, vii, 352; Holbrook, 1842, *54*, iii, 139, pl. 20; Boulenger, 1885, *29*, ii, 281; *Opheosaurus ventralis*, Cope, 1875, *12*, 46.

Long, slender and snake-like. Head small, narrow, not separated from the body by a distinct neck. Snout long and tapering to the rounded tip. Upper surface of the head with a large vertical plate. Behind this is the interparietal, with a minute whitish spot, the "pineal eye." Behind this plate is the small occipital. Ear opening in a line with the mouth; larger than the nostril. Palatine teeth present. The pterygoid teeth in three to five longitudinal series. Teeth all conical.

On the dorsal surface, between the lateral folds, are 14 to 16 longitudinal rows of scales; between the folds across the lower surface are 10 rows; between the head and the vent are about 120 transverse rows. All the scales are smooth, except those of the median dorsal row, which are obtusely keeled. The tail is about twice as long as the head and body.

The color of the upper surface and the sides is olive, with several longitudinal narrow lines of brown. In life the lighter colors are green and yellow. There may be a dorsal dark band with a broad olive band on each side, followed on the sides by three dark lines above and two below the lateral fold; or the broad bands of olive may contain two additional narrow lines, in which case the whole upper surface and the sides are narrowly streaked with brown and yellowish or green. A specimen is occasionally found which has the upper surface black, without any lines, and the sides with rows of white ocellated spots. Again, the lines may be wholly replaced by rows of spots. The abdomen is yellow, sometimes orange, occasionally slate color. Head brownish or olive, with mottlings of darker. Lower jaw yellow.

The average length is about 18 inches, but specimens may occur as much as 3 feet.

The distribution is from Southern Virginia west to Wisconsin and south to Mexico.

In Indiana this reptile has not shown itself to be widely distributed or abundant. Professor John Collett tells me that he has seen it in Warren county. I have a specimen that was secured at Wolcott, in White county, by Mr. Charles S. Beachler. I find mention (7, 30, 122) of one being taken in Starke county. Mr. Robert Ridgway informs me that it was formerly a very common species in the Lower Wabash Valley, especially about Mt. Carmel. Since the reptile is most abundant in the Southern States, this finding of it in the most northern portion of Indiana indicates that we are likely to discover that it is a resident of the whole State.

On account of the serpent-like form of this lizard, it is almost universally regarded as a snake. It may be distinguished from the members of the order of serpents by the little distensible mouth, the firm union of the sides of the lower jaws at their symphysis, by the possession of eyelids, and by the rows of small scales covering the belly. On the lower surface of snakes there is a series of plates that pass across from one side to the other.

Wherever this lizard is found it has attracted a great amount of attention, especially on account of the facility with which the whole animal appears to break up into short pieces. Along with this knowledge of its fragility, which has given it the name of "Glass-snake," there goes the tradition that these pieces have the power of reuniting themselves, so that the reptile is thoroughly reconstructed and as sound as ever. Concerning these matters there has been a great amount of discussion in the newspapers and the scientific journals. As regards the liability of the animal to break up into pieces of different lengths on being struck or roughly handled, there is no doubt that the popular notion is correct. Two-thirds or more of the Glass-snake is tail. It is a well-known fact that many lizards on being seized drop their tails in order to free themselves or to deceive the pursuer. The tail thus lost may be reproduced. When occasion appears to demand the sacrifice, the Glass-snake sunders its tail into a number of wriggling pieces, and while the astonished observer stands viewing the wreck, the head and body hastens to a place of safety. In order that all these pieces might unite again to form a sound lizard, they would have to be fitted together in the proper order, and with the ends turned in the right direction; the half dozen or more conical muscular masses which project from the ends of the pieces would have to be interdigitated accurately; the nerves and blood vessels would need to come into juxtaposition; and then all the torn surfaces unite by "immediate union" so quickly and effectively that the animal can betake itself to its business. Hard as the possibility of accepting all this is for the naturalist, such is the popular notion, and a writer in the Scientific American, September 3, 1887, says that he seen the thing done. Before scientific men will believe the assertion it will have to be well corroborated.

The specimen of the Glass-snake which was taken in Starke county, Ind., was sent to the editor of the Popular Science Monthly (vol. xxx, p. 122,) for examination. The account of it is mixed up with the name of the Chain-snake, but the reference to the position of the vent shows plainly that it was a Glass-snake, as the editor also says it was.

This animal appears to select for its abode dry, rather than damp, situations. Holbrook states that it often is found in sweet-potato hills in time of harvest. It is entirely harmless, and can be easily tamed. It probably will not bite at all, and could do no injury if it did; but people who know little about reptiles think that all snakes and lizards, as well as all salamanders, are poisonous. It is said that the Glass-snake appears very early in the spring, even before the true snakes, and remains late in the autumn. About its breeding habits I have been able to learn nothing definite. It probably lays its eggs in the ground. Drs. Coues and Yarrow say that several individuals of this lizard were eaten by a specimen of *Ophibolus getulus sayi* that was kept in the same cage.

Family V. TEIIDÆ.

Lizards with form varying from moderately stout to vermiform. Limbs usually present and well developed; sometimes rudimentary, and in one genus the hinder limbs are absent. Dentition various. Tongue thin, flat, ending in two long smooth points, and having its upper surface covered with overlapping scale-like papillæ. Plates of the head large or small. No bony plates underlying the epidermal scales.

A family comprising 35 genera and over 100 species; all living in the New World.

Genus CNEMIDOPHORUS, Wagler.

Cnemidophorus, Wagler, 1830, 75, 154; Boulenger, 1885, 29, ii, 360. Body terete, head pyramidal, tail long and slender. Limbs well developed, with five digits each. Head with large shields. The anterior nasals meeting in the middle line. Eyelids present. Tympanum exposed, sunken below the surface. Lateral teeth bicuspid or tricuspid. Scales of the upper surface of the body small; those of the abdomen large, and arranged in 8 to 12 longitudinal rows. A double fold of skin on front of neck. Femoral pores present.

About 16 species are known.

Cnemidophorus sex-lineatus, (Linn.).
Six-lined Lizard.

Lacerta sexlineata, Linnæus, 1766, 64, ed. xii, i, 364; *Ameiva sexlineata*, Holbrook, 1842, 54, ii, 100, pl. 15; *Cnemidophorus sex-lineatus*, Boulenger 1885, 29, ii, 364.

Long and slender. Head a four-sided pyramid. A large vertical plate, on each side of which, over the eyes, are four supraoculars. Nostril in the anterior nasal. Ear about the size of the eye. A few pterygoid teeth present. Gular folds two, the posterior with large scales in front. Upper surface of the fore limbs with large scales. Front of thigh and under surface of the lower leg with enlarged scales. Scales of the upper surface of the body small; those of the abdomen large, squarish; and arranged in 8 longitudinal, and about 33 transverse rows. Tail with large, squarish, keeled scales, the keel running somewhat obliquely. Vent with three enlarged scales in front. Appressed hind limb reaching to the ear. Tail nearly twice the length of the head and body; sometimes more.

The color of the back is olive or brownish, in life green. The sides are black, with three narrow stripes of yellow. A greenish stripe on the back of the thigh, prolonged on the sides of the tail.

Total length about 10 inches. Distributed from Maryland to Kansas, Southern California, and Mexico.

This species is included in the fauna of Indiana on the testimony of Mr. Robert Ridgway of the National Museum. He writes me that while he was collecting snakes at Monteur's Pond, near Wheatland, Knox county, he climbed a buttonbush to snare a big water-snake. While maneuvering to get his noose-rod into position he saw a specimen of what he is positive was this species (with the appearance of which he was previously very familiar) basking in the sun on a branch of the same bush. Before he could secure the lizard, it jumped off into the water and swam away. There is nothing, whatever, improbable about the occurrence of this lizard in Southern Indiana, since I have taken it in Western Illinois, at a point as far north as Logansport, Ind. Since the above was written Mr. A. W. Butler has informed me that he found this species in a small collection of reptiles sent him from Bloomington, Ind. I have not seen the specimen.

HABITS—Probably the most remarkable quality possessed by this animal is its ability to run with great swiftness. Dr. Coues in speaking of it says that the eye can scarcely follow it when it is running at its best on level ground. Dr. Yarrow also states that it is very difficult to capture, since it runs with the greatest celerity over the sand and rocks. In order to capture it Dr. Coues used fine shot in a horse-pistol. It may be taken in a butterfly net. It lives in high and dry situations, among bushes, brush heaps, logs, and the like. The naturalists named above do not regard it as at all arboreal in its habits; still it may occasionally resort to trees. Its food is principally insects. Dr. Coues states that it would come into the tents and catch flies in a quiet, furtive way, but on the least alarm it would dart out of sight. At Ft. Macon, N. C., (*1*, '76, 47,) it was seen early in April, and could be found until the cold weather in October. Notwithstanding its agility, it is one of the animals oftenest found in the stomachs of the larger snakes. Holbrook says that he has often seen the male and the female toward evening hunting in company for insects. I know nothing about its breeding habits.

Family VI. SCINCIDÆ.

Lizards from short and stout to long and worm-like. Limbs present or absent. Dentition pleurodont; form of teeth various. Tongue thin, flat, moderately long, free, and slightly notched in front; covered with overlapping scale-like papillæ. Head with symmetrical plates. Epidermal scales supported by bony plates. Ovoviviparous.

Cosmopolitan. Contains 25 genera and about 400 species. We have 2 genera, each with a single species.

Palate cleft by a single median slit. *Lygosoma*, p. 138.
Palate with two slits, one from each nostril. *Eumeces*, p. 139.

Genus LYGOSOMA, Gray.

Lygosoma, Gray, 1828, *110*, iii, 228; Boulenger, 1887, *29*, iii, 209; *Oligosoma*, Girard, 1852, *5*, 196.

Body short and stout or elongated. Limbs well developed, rudimentary or absent. Outer nostril in the nasal plate. Eyelids well developed. Tympanum distinct or hidden. Maxillary teeth conical or obtuse. Pterygoid teeth minute or absent. Palatine bones in contact, so as to leave a single median slit from the internal nostrils to opposite the middle of the eyes.

A widely distributed genus containing more than 150 species.

Lygosoma laterale, (Say).

Brown-backed Lizard.

Scincus laterale, Say, 1823, *14*, ii, 324; *Lygosoma laterale*, Holbrook, 1842, *54*, ii, 133, pl. 19; Boulenger, 1887, *29*, iii, 263; *Oligosoma laterale*, Cope, 1875, *12*, 39.

A small lizard with a short head, elongated body, a long terete tail, and feebly developed limbs. Head flat above, the perpendicular sides tapering to the snout. Lower eyelid with a transparent disk. Ear opening the size of the opening of the eye, the tympanic disk deeply sunken. No gular folds. Limbs lacking much of meeting along the side, the appressed hind leg reaching hardly a third of the distance to the axilla. Body covered with smooth scales, not arranged in straight rows across the body. About thirty scales in oblique row around the middle of the body.

The color above is olive or brownish, with scattered brown spots, the tail often blue. Side with a brown, white margined stripe running from the eye on to the tail. Below this may be a similar but narrower stripe passing between the fore and the hind limb. Abdomen greenish white.

The total length is about 5 inches, sometimes as great as 6 or 8 inches. The tail may make up as much as two-thirds of this. The species ranges from Florida and Texas to North Carolina, Indiana and Kansas. Indiana localities from which this species has been reported are few, but it probably inhabits much of the southern portion of the State. Mr. Robert Ridgway has taken it at Wheatland, and has seen it in the "watermelon sands" district of Gibson county.

HABITS.—In some portions of the Southern States this species of lizard is abundant. Holbrook says that they might, in his day, be seen by thousands in the thick forests of oak and hickory of Carolina and Georgia. They emerge from their retreats, after sunset, in search of small insects and worms, on which they live; yet they appear and disappear so rapidly that they might at first be mistaken for crickets or other insects. It is difficult to secure them alive. They conceal themselves rapidly under roots or beneath leaves. Baird and Girard (*37*, 229), say that it is always met with running on the surface of the ground in forests, among dead leaves, never ascending either trees or shrubs like many other lizards. It is said to breed in Georgia in March.

Genus EUMECES, Wiegmann.

Eumeces, Wiegmann, 1835, *83*, ii, 288; Boulenger, 1887, *29*, iii, 365.

Lizards of moderate proportions. Limbs well developed. Outer nostril in the nasal plate. Eyelids with scales. Tympanic disk distinct, deeply sunken. Maxillary teeth with conical or rounded crowns. Pterygoid teeth present. Palatine bones not meeting in the median line; palate therefore with two clefts, one from each nostril.

Of this genus 31 species are recognized. N. A., Cent. A., Asia, Africa.

Eumeces fasciatus, (Linn.).

Red-headed Lizard; Blue-tailed Skink.

Lacerta fasciata, Linnæus, 1758, *64*, ed. x, i, 209; ed. xii, 369; *L. quinquelineata*, Linnæus, 1766, *64*, ed. xii, 366; *Plestiodon erythrocephalus*, Holbrook, 1842, *54*, ii, 117, pl. 16; *Scincus quinquelineatus*, Holbrook, 1842, *54*, ii, 121, pl. 17; *Scincus fasciatus*, Holbrook, 1842, *54*, ii, 127, pl. 18; *Eumeces fasciatus*, Cope, 1875, *12*, 45; *Eumeces quinquelineatus*, Boulenger, 1887, *29*, iii, 369.

Head short, broad behind, wedged-shaped in front. Snout short, obtuse. Nasal plate pierced by the nostril. Behind this is the postnasal, which joins the first labial. Behind the postnasal comes the loral, which joins the plate in the median line, the fronto-nasal. Vertical large, in contact with three anterior supraoculars. On the back of the head, in the median line, is the interparietal, in which a faint whitish spot indicates the position of the pineal eye. Ear opening smaller than the eye, the tympanic disk deeply sunken. No distinct neck nor gular fold. When the limbs are pressed to the sides they overlap. Tail longer than the head and body. Length of hind limb in the distance from the snout to the vent nearly or quite two and a half times. Scales of the back, sides and abdomen about equal in size, smooth, 28 or 32 longitudinal rows around the middle of the body. A row of enlarged scales on the under side of the tail.

Young and middle sized individuals are nearly black above, with five yellow lines running from the head to the middle of the tail. The median line forks on the head. The extremity of the tail is often bright blue. The abdomen is bluish white. As the animal grows older the stripes become obscure, the general color fades to olive or brownish, and the head becomes bright red. These older specimens were long regarded as belonging to a distinct species. Even Holbrook, who lived where the animal is abundant, described it under three names, as cited above.

The total length may be as much as 13 inches, and even more.

The species is distributed from Massachusetts to Wyoming and south. In Indiana it appears to be pretty generally and abundantly distributed in the southern half of the State. Since it has also been found in Michigan, there seems to be no reason why it should not also occur in Northern Indiana. The following are localities from which I have seen specimens:

Brookville (Hughes); New Harmony (Sampson's coll.); Monroe county (Bollman and Nor. Sch. coll.); Terre Haute (Blatchley); Marion and Crawford counties (Hay); Wabash county (A. B. Ulrey). A specimen from Bloomington is 10 inches long.

HABITS.—This is a very active species of lizard, and is found in forest regions, living about old logs and stumps, and under bark. It often ascends trees, and it would appear that the older individuals spend the greater part of their time there. Holbrook says of these older ones, called by him *P. erythrocephalus*: "They choose their residence in deep forests, and are commonly found about hollow trees, often at a height of thirty or forty feet from the ground; sometimes in a last year's nest of the woodpecker, out of which he thrusts his bright red head in a threatening manner to those who would disturb his home. He never makes his habitat on or near the ground, and, in fact, seldom descends from his elevation unless in search of food or water. Though shy and timid, he is very fierce when taken, and bites severely." The bite is not poisonous. Of the less mature individuals, Holbrook says that they live mostly on the ground, in holes of stumps, under the bark of decaying trees, and similar places. They seldom ascend trees unless they are pursued. Their food consists of insects. Smith (*18*, 651), states that this animal lays nine oval eggs at a time.

Dr. Abbott (7, 34, 170), thinks that the Blue-tailed Skink has a greater degree of intelligence than the Pine-tree Lizard, *Sceloporus undulatus*. The latter, on being captured, immediately submits and becomes tame. The Skink, on the other hand, is fierce, and defends itself with vigor. One that he captured and put into a case immediately buried itself and scarcely made an appearance for a week. After it came from its retreat it was suspicious of every object and sound. When it went into its retreat it never returned by the same burrow, but would stick its head out at another hole and carefully study the outlook. At length it gained sufficient confidence to eat, and in about four months it became comparatively tame. Its disposition is, in many respects, the opposite of that of the Pine-tree Lizard.

Dr. J. Schneck (*22*, 14, 55), mentions some specimens of this species which had the tail forked, as the result probably of a loss and reproduction of that appendage. When these lizards are pursued and caught by the tail, this is very liable to be dropped. The tail is afterward reproduced, and it seems that occasionally two tails take the place of the lost one. The same observer speaks of the great variations in the color of this species due to differences in age, and of the aged ones as "ruby-headed and copper-bellied specimens, 20 inches in length."

Order CHELONIA.

Tortoises and Turtles.

Reptiles having the trunk relatively short and broad, with the upper and the lower walls forming two disks, which are united on the sides between the fore and hind limbs. The disks strengthened by bony deposits, which usually, but not always, involve the endo-skeleton. The bony dorsal shield (carapace) usually formed of the expanded and suturally united ribs and vertebral spinous processes. The lower shield (plastron) composed of the clavicles and a few dermal bones. No true sternum. Trunk rigid; only the neck and tail flexible. Jaws without teeth and covered with horny sheaths. Eyes with lids and a nictitating membrane. Tympanic membrane external, sometimes hidden by the skin. Tongue thick and fleshy. Limbs four; developed for walking, except in the marine turtles, in which they are formed for rapid swimming. Reproduction by means of eggs, spherical or elliptical in shape and protected by a calcareous shell.

Upper and lower disks without horny epidermal plates. A soft skin.
Trionychoidea, p. 142.
Shell covered with large, symmetrical horny epidermal plates.
Testudinata.

Suborder TRIONYCHOIDEA.

The dorsal vertebræ and the expanded ribs involved in the carapace. Plastron composed of 9 bones, which enclose a fontanelle. The dorsal disk rarely strengthened by marginal bones, its border therefore flexible. The horny sheaths of the jaws hidden by fleshy lips. Fourth digit with 4 or more phalanges.

A single family living in the rivers of North America, Asia and Africa.

Family VII. TRIONYCHIDÆ.

Body broad and much depressed; the margins of the carapace thin and leathery, in rare cases having marginal bones. No epidermic scutes. Snout much produced, leathery, with the nostrils at the tip. Ear hidden. Only the three inner digits furnished with claws. Head and neck completely retractile.

Of this family 6 genera are recognized, only one of which lives in America.

Genus TRIONYX, Geoff.

Trionyx, Geoffroy, 1809, *85*, 84; Gray, 1855, *25*, 64; Boulenger, 1889, *84*, 242.

Plastron little developed behind, leaving the hinder limbs and tail completely exposed; with not more than 5 callosities No marginal bones in the border of the carapace. Jaws strong.

A genus embracing about 15 species, residents in the rivers of North America, Asia and Africa.

A. Nostril circular, having no papilla projecting into it from the septum. (*Amyda*, Agassiz.) *T. muticus*, p. 143.
AA. Nostril crescent-shaped, having a papilla projecting into it from the septum. (*Aspidonectes*, Agassiz.)
    a. Under side of feet white, not mottled with brown.
                                        T. agassizii, p. 144.
    aa. Under side of feet mottled with white and brown.
                                        T. spiniferus, p. 146.

Trionyx muticus, LeS.

Spineless Soft-Shelled Turtle

Trionyx muticus, LeSueur, 1827, *86*, 263, pl. 7; Holbrook, 1842, *54*, ii, 19. pl. 2; Gray, 1855, *25*, i, 69; Boulenger, 1889, *84*, 260.

Amyda mutica, Agassiz, 1857, *4*, i, 399; *Aspidonectes muticus*, Baur, 1888, *22*, 1122.

Head long, low, and pointed in front, descending rapidly in front of the eyes. Skull, from the eyes forward, drawn out, the margins of the upper and the lower jaw being concave outwardly. The horny upper jaw with a cutting edge, which is deepest forward and bluntly toothed posteriorly. Lower jaw also with a sharp edge, and both jaws furnished with an alveolar surface, the leathery snout ending obliquely, so that the nostrils are somewhat under the tip. The nostrils are circular, there being no papilla projecting into them from the septum.

Body flat and oval. No trace of keel along the middle of the back; often a depression instead. No spines along the anterior border of the carapace, nor any tubercles anywhere. Callosities well developed on the plastron of the adults, especially of the males.

The color above is brownish, olive, or bluish-gray. In the young there are some blotches of dark brown. On the margins of the carapace, laterally and posteriorly, is a band of yellow, bordered internally with black. These bands are likely to disappear later in life. Head with a white stripe, margined with black, from the eye over the ear and then descending on the neck. Head and neck below the level of the edge of the upper lip white, without any mottling. Under surface of the feet white or bluish-gray; never mottled as in *T. spiniferus*.

The length of the carapace of fully grown adults may be a foot, sometimes probably more. The females have tails that scarcely project beyond the edge of the carapace, while that of the male is much longer. Both Agassiz and Baur have noted the fact that the males are much smaller than the females. Baur has also made the observation that the males are not so numerous as the females (*22*, 22, 1122). It is not at all improbable that such is the case; yet it may be only apparently so, due to fewer of the males being captured, on account of their smaller size, or on account of the saving only of the larger, finer specimens.

The species inhabits North America from the St. Lawrence River to Florida and west to the plains. I have note of its occurrence in the rivers of Indiana at five points: Delphi (Agassiz); Madison (Yarrow); Mt. Carmel, Ill. (R. Ridgway); and Terre Haute (Blatchley). LeSueur described the species from specimens taken at New Harmony.

HABITS.—This species, like all the soft shelled turtles, is wholly aquatic, since they leave the water only on rare occasions. They delight to remain about the roots of trees which have fallen into the water or in drifts of timber. Here they can watch for prey and not be observed by any supposed enemy. Away from such means of concealment they are accustomed to bury themselves completely in the sand, leaving only their heads exposed. Since their heads do not differ much in color from the sand it is difficult for one to recognize them, even when the eye is directed to them. When air is required it is obtained by stretching out the neck until the pointed snout reaches the surface. The head is then again withdrawn. Like *T. spiniferus*, this species no doubt enjoys a true aquatic respiration. They subsist probably on insects, fishes, water snails, and similar small animals. Agassiz found the larvæ of neuropterous insects in their stomachs. Max. Von Wied (*103*, xxii, 53) says that LeSueur found in their stomachs worms, snails, fruits, and even hard nuts. If there are potatoes growing near the water the turtles find their way to them and devour the stems, of which they are very fond.

The eggs are spherical in form, about seven-eights of an inch in diameter, and have a thick, but brittle, calcareous shell. They are deposited in the sand on the shores of the rivers where the adults live. The young are flatter and more nearly circular than the adults. This species, like the other species of *Trionyx*, is regarded as a great delicacy.

Trionyx agassizii, (Baur).

Agassiz' Soft-Shelled Turtle.

Trionyx ferox, in part, Boulenger, 1889, *84*, 259; *Platypeltis ferox*, Agassiz, 1857, *4*, i, 400, and ii, pl. vi, fig. 3 (young), pl. vii, fig. 22 (egg); *Platypeltis agassizii*, Baur, 1888, *22*, xxii, 1121.

Head with the fore part not drawn out, the margins of the jaws not concave outwardly. Snout with the nostrils at the tip; the latter crescent shaped, there being a papilla projecting into it from the septum. Body with a low, obtuse keel along the middle line. Spines present along the anterior border of the carapace, these largest in the males. Upper surface of the carapace often rough with minute tubercles. Callosities of the middle and hinder part of the plastron well developed.

Color of the upper surface olive, with blotches or spots of black. A light streak, margined with black, starts at the tip of the proboscis, divides just in front of the eyes, and sends a branch through each eye to the side of the neck. The younger individuals have a yellow border around the lateral and posterior edges of the carapace, inside of which are two or three lines of black. The spots on the carapace of the young are solid, but later in life they may become ocellated. The lower surfaces are white. The bottoms of the feet are said by Agassiz to be always free from mottlings of black. In a specimen which came from Mississippi, however, the under surfaces of these are moderately mottled with black. Agassiz states that the largest individual of this species of which he had any knowledge was 18.5 inches long from the front to the hinder end of the carapace. They are usually much smaller. The tail of the male projects beyond the edge of the carapace; that of the female does not.

This species closely resembles *T. spiniferus*. It is to be distinguished by the solid spots and the two or three black lines around the carapace (when not too large), the light line of the head dividing behind the base of the proboscis, and the uniformly colored lower surfaces of the feet.

This species belongs to the Southern States from South Carolina to Texas. A single specimen has been forwarded to the National Museum from Madison, Ind. It ought to be sought for all along the Ohio River.

Whether or not this species is more vicious than its relatives can hardly be said. But all the American species are ready to snap and bite whenever they are teased, and their biting is not to be held in contempt. The head and long neck can be thrust out with great rapidity, and the sharp edged jaws are like scissors. Holbrook says that it will sometimes leap up and give a loud hiss. He further states that it is very voracious, feeding on fish and such reptiles as it can secure, and is so greedy that it takes the hook readily when baited with any substance whatever. Yet he had never known them to take food in captivity, even after several months. They swim with great rapidity, and often conceal themselves in the mud, buried two or three inches deep, leaving only a small breathing hole for the long neck and small head. This it occasionally thrusts out, but usually keeps it concealed so that a passer-by might think the hole the residence of some large insect. They are often seen basking in the sun on rocks and apparently asleep.

In the South they lay their eggs in May. These are about 60 in number, have a thick, smooth, brittle shell, and are larger than those of *T. muticus*, being a little less than an inch in diameter. They are hidden in the sand along the shores of streams.

Dr. Baur considers the form found in the Mississippi Valley as an entirely distinct species from the *Testudo ferox* of Schneider. Should he be correct in this judgment, as he probably is, the species above described must be known as *Trionyx (Platypeltis) agassizii*.

Trionyx spiniferus, LeS.

Spiny Soft-Shelled Turtle.

Trionyx spiniferus, LeSueur, 1827, *86*, 258, pl. vi; *Trionyx spinifer*, Boulenger, 1889, *84*, 259; *Aspidonectes spinifer*, Agassiz, 1857, *4*, i, 403, and ii, pl. vi, fig. 1 (young), pl. vii, fig. 23 (egg).

Resembling much *T. agassizii*. Skull tapering gradually to the snout. Proboscis with the nostrils at the tip; these crescentic in shape, a papilla projecting into each from the septum. A low obtuse keel along the middle of the back. A series of spines on the front edge of the carapace, largest in the females. Whole upper surface of carapace often covered with minute asperities, also more prominent in the females. Tail of the male projecting considerably beyond the carapace. Callosities well developed in the middle and hinder parts of the plastron. General color above olive or light brown. In the young there are numerous ocellated spots, or rings, of black all over the carapace. These may be retained until the size has become considerable, but they finally become irregular blotches. In the young and half grown there is a yellow border around the sides and posterior edge of the carapace, and just within the yellow border is a single line of black. Head olive, with a light line, margined above and below with black, starting at the tip of the proboscis, forking at its base, and sending a branch through each eye and down on the neck. The plastron is white. Under surface of the feet much mottled with white and black.

About the size of *T. agassizii*. Agassiz states that the largest of which he had knowledge had a carapace 14 inches long.

This species is to be distinguished from *T. agassizii* by the mottled lower surfaces of the feet, the line of the head, *forking at the base of the proboscis*, and, in case the specimen is not too old, the ocellated spots of the carapace, and the single dark line around the edge of the carapace.

Habitat from Vermont to Montana and south to the Gulf. More abundant in the Northern States. In Indiana it is the most common species of soft-shelled turtle, and is so generally distributed throughout the State that it is not necessary to mention localities.

Habits.—The habits of this turtle are much like those of *T. agassizii*. It lives in similar localities, captures the same kinds of food, and deposits its eggs in the sand, just as *agassizii* does. One was found by myself on the 14th of March buried in the sand where the water was so deep that she could only with difficulty reach the surface with her proboscis. The head, colored just like the sand, was drawn entirely under as soon as she saw that she was observed. It was with a good deal of exertion that she was dislodged. She was kept until the 13th of May, during which time she could not be induced to eat anything. On being put into a ditch she immediately buried herself in the mud, and when hidden she gave her body some sidewise movements so that the mud settled over her as though nothing had disturbed it. Here she remained a day and two nights. Prof. Blatchley (*94*, '91, 34), states that he has seen them moving freely about in the water as late as December 11, and as early as March 19. When hibernating they burrow in the mud at the bottoms of ponds and streams.

This specimen will bite severely, as several observers have had opportunity to learn. DeKay mentions the fact that one bit a dog and took out a mouthful of hair. These turtles can run rapidly on the land, and when in the water they swim with great swiftness, as any one knows who has tried to catch them. Profs. S. P. and S. H. Gage have demonstrated that both this species and *T. muticus* enjoy a true aquatic respiration. They say that these animals often remain voluntarily under the water for from two to ten hours consecutively. While under the water there are about 16 movements of the hyoid apparatus each minute, and by means of these the mouth and pharynx are filled with water and again emptied. The mucous membrane of the pharynx is closely beset with filamentous processes which have the appearance of the villi of the intestines, and are abundantly supplied with blood. Analysis of the water in which a turtle was kept some hours proved that it was deprived of its oxygen and filled with carbonic acid.

The number of eggs laid by this species is probably about the same as for *T. agassizii*. The eggs are spherical, have a thick, brittle calcareous shell, and under this a very tough membrane. The eggs are a little larger than those of *T. agassizii*, an inch in diameter. LeSueur (*86*, xv, 263) says that at New Harmony the females lay their eggs in April and May in the sand along the river bank. He has found in them 50 or 60 eggs, about 20 of which were ready to be laid. The others were probably the eggs of the next season. The young appear in August. This turtle is highly prized as an article of food.

Suborder TESTUDINATA.

Carapace usually very complete, formed by the expanded spinous processes of the vertebræ, the expanded ribs, and a series of dermal marginal bones. Plastron consisting of 8 to 11 bones; commonly united by suture with the carapace. Both carapace and plastron, with one exception, covered with large, symmetrically arranged epidermal plates. Jaws covered with horny sheaths; not hidden by fleshy lips. Fourth digit never with more than 3 phalanges.

KEY TO THE FAMILIES OF TESTUDINATA.

Plastron narrow, cross-shaped; bridge long and narrow; marginals 23, not including the nuchal; tail long. *Chelydridæ*, p. 148.

Plastron of small to moderate size; bridge short, but wider; marginals 23; tail short. *Kinosternidæ*, p. 150.

Plastron filling the openings of carapace, or nearly so; marginal plates 25; tail short to moderate. *Testudinidæ*, p. 155.

Family VIII. CHELYDRIDÆ.

Body broad and depressed; the shell highest in front, serrated along its posterior border. Plastron formed of 9 bones; small and cross-shaped; the bridge narrow. Abdominal scutes separated from those of plastron by a series of inframarginals. Head large, jaws strong and hooked. Tail long, with one or more rows of compressed, horny tubercles above. Eggs spherical.

Genera two, both of which are represented in Indiana.

Shell without additional plates between the marginals and costals.
Chelydra, p. 148.

Shell with 3 or 4 extra plates between the marginals and costals.
Macrocemys, p. 151.

Genus CHELYDRA, Schweigg.

Chelydra, Schweigger, 1814, *88*, 23; Agassiz, 1857, *4*, 1, 416; Boulenger, 1889, *84*, 20.

Carapace with three tuberculated keels, which disappear more or less with advanced age. No supernumerary scutes intervening between the marginals and the costals, just above the bridge. Plastron small, with five pairs of scutes, the abdominals apparently displaced and covering the bridge. Head large, jaws hooked. Head with the skin marked off into somewhat symmetrical plates. Tail with two rows of large scales beneath.

Chelydra serpentina, (Linn.).

Snapper; Snapping Turtle.

Testudo serpentina, Linnæus, 1758, *64,* ed. x, 199; *Chelydra serpentina,* Schweigger, 1814, *SS,* 24; Agassiz, 1857, *4,* i, 417, and ii, pl. iv, figs. 13-16, pl. vii, figs. 24-26; Boulenger, 1889, *84,* 20, with figs. *Chelonura serpentina,* Holbrook, 1842, *54,* i, 139, pl. 23.

Carapace broad and rather depressed, highest in front and notched behind. A median and two lateral, tuberculated keels, disappearing late in life. Marginal plates, exclusive of nuchal, 23. Vertebral scutes wider than long, tuberculated behind. Costal scutes tuberculated near the upper posterior angle. Plastron small, leaving the limbs exposed; covered with five pairs of scutes; the bridge very narrow. Two or three inframarginals at the outer end of the bridge.

Head large and flattened above, with rather conspicuous bony ridges; tapering, but not descending toward the snout. Eyes directed upward and outward. Feet broad and webbed to the nails. Fingers five, all with nails. Toes five, the outer one without a nail. The outer border of all the limbs with a sharp fold of skin which greatly increases the surface of the limb, as aid in swimming. Tail long and pointed, equaling the length of the plastron. Tail furnished above with a median row of large horny tubercles, supported by a bony core. Each side of the tail with smaller tubercles. Under side of tail with two rows of large scales.

Skin of neck, under jaw, body, limbs and tail covered with wrinkles and large and small warts. Fore-arm and hands and feet with large, overlapping, sharp-edged scales. Color of the carapace chestnut brown to black. Plastron and soft skin whitish or yellow. Head and upper neck brown. Attains a total length, it is said, of four and a half feet, the shell two feet, usually much smaller. The weight may reach from 20 to 30 pounds.

This species has a remarkably wide distribution. It occurs from Nova Scotia to Ecuador, in South America. Westward it probably extends to the Rocky Mountains. It is found, no doubt, in every stream and pond in the State of Indiana.

HABITS.—This turtle spends the greater portion of its life in, or closely about, streams and lakes and ponds. Although found living in clear rivers, it appears to prefer muddy ponds. It is often seen far away from any water, walking along with awkward and halting gait. Its mode of locomotion has been compared with that of the alligator. When seen on the land it may be seeking some spot in which to deposit its eggs, or seeking for food, or perhaps crossing from one stream to another. In the water they do not seem to swim, but they may often be seen floating along just below the surface, with the snout and eyes only exposed. When disturbed, they immediately go to the bottom, and conceal themselves there. When traveling about they are often seen with a great amount of mud on their backs as though they had been burrowing in the earth. The Snapping-turtle is strong and courageous. When attacked they neither attempt to retreat nor retire passively into their shells, as do most turtles. The jaws are opened, the head and long neck are suddenly thrust out, and at the same moment the animal leaps forward toward its tormentor. If the aim has been correct, the jaws close on the enemy and the hold is doggedly retained. It is a curious notion held by many people that, when it has once secured a hold, it will not let loose until it has thundered. It will sometimes permit itself to be carried around by a stick which it has seized.

The Snapping-turtle is wholly carnivorous and extremely voracious. Their food consists of frogs, fishes, the smaller and younger water fowl, and crayfishes. They do not hesitate to eat any animal substance that presents itself. They have been accused of capturing young ducks. A large specimen that I dissected had in its large intestine the feathers and partially digested bones of a full grown robin. The wing and tail feathers filled up the intestines. Its excrement contained the remains of a crayfish. I have been told that they will steal the sportsman's string of fish, and use the forefoot in tearing off what they can not get into the mouth.

The eggs are laid during the month of June, and hatch in the autumn. They number from 30 to 70, and are deposited in holes excavated along the banks of streams. Agassiz says that the hole is excavated at first directly downward and then laterally, so that the eggs are deposited on one side of the mouth of the excavation. They are all deposited in one hole. After the eggs are laid the female covers them up, smooths the sand over them, and leaves them to their fate. The eggs are spherical, about an inch in diameter, and provided with a calcareous shell. The shell is not brittle, but somewhat less flexible than that of most tortoises. Occasionally an elliptical egg is found. Agassiz is authority for the statement that the young will snap before they have escaped naturally from the egg.

The flesh of the Snapping-turtle is often used for food, especially that of the younger individuals. When they grow old their flesh is likely to have a musky and disagreeable smell. Mr. True states that these turtles are regularly seen every spring in the markets of Washington ready for cooking. Storer wrote that in Massachusetts many persons saved the oil of this animal and used it to heal bruises and sprains. As a therapeutical agent it is worthy to stand alongside of goose oil, skunk oil, and rattlesnake oil.

Genus MACROCLEMYS, Gray.

Macroclemys, Gray, 1855, 25, 48; *Macrochelys*, Gray, 1855, 25, sup., 64; *Gypochelys*, Agassiz, 1857, 4, i, 413; *Macroclemmys*, Boulenger, 1889, 84, 23.

Carapace with three prominent keels, which persist throughout life. A series of three or four supernumerary marginal scales on each side, between the normal marginals and the costal scutes. Plastron small, cross-like, and with five pairs of scutes. Bridge narrow. Head very large, covered with smooth, symmetrical plates. Orbits looking outward and forward. Jaws very strong and hooked. Tail with three series of tubercles above; the lower surface with small scales.

Macroclemys temminckii, (Troost).

Alligator Snapping Turtle.

Chelonura temminckii, Troost, 1842, 54, ii, 47, pl. 24; *Macroclemys temminckii*, Gray, 1855, 25, 49; *Macroclemmys temminckii*, Boulenger, 1889, 84, 25, with figures; *Macrochelys lacertina*, Cope, 1872, 1, 23; *Gypochelys*, Agassiz, 1857, 4, i, 414, and ii, pl. v, figs. 23-27.

Carapace furnished with three prominent keels which do not vanish with age. Each median scute rises posteriorly into a knob, which is largest on the hindermost vertebral scute. The lateral keel is located on the upper ends of the costal scutes. The keel rises on the hinder border of each scute into a knob, and the knobs on the hinder scutes are the highest. Posterior border of the carapace serrated. Between the lower ends of the anterior three costal scutes and the marginals occur three or four supramarginals. The plastron resembles that of *Chelydra serpentina*. The head is of enormous size, broad behind, tapering rapidly to the acuminate beak and snout. Beak of upper jaw projecting beyond the lower, and strongly hooked, the outline of the cutting edge rising from the point of the beak, then descending to the middle, and then rising to the corner of the mouth. Lower jaw turned up into a strong hook. Head covered with large, symmetrical plates. Neck short. Tail about three-fourths the length of the carapace, furnished above with three rows of low tubercles, below with rows of small scales. Color yellowish or reddish brown to black.

This species attains a great size for a fresh-water turtle. Agassiz saw one alive that weighed about two hundred pounds. One of his correspondents speaks of a skull which measured nine inches between the eyes. A dry specimen that I examined in the National Museum had the carapace 23 inches long and 20 wide. The head was 8 inches long and 7 wide. The sternum was 16 inches long. A skull in the Indiana

Geological Museum, said to have come from Arkansas, is 9 inches wide and as many long. The length from the snout to the occipital condyle is 7¼ inches.

Its range is from western Georgia to Texas and north to Indiana. It has been taken in the Wabash River, at Grayville, Ill., as Mr. Robert Ridgway, of the National Museum, informs me. The specimen captured there was exhibited at county fairs, and was so strong that it could easily walk about with a large man on its back. Dr. Yarrow (*10*, 30) reports two specimens of this species in the National Museum, from Northville, Mich., but an examination of the records at the Museum shows that the specimens sent from that place were not of this tortoise, but of *Necturus*.

Habits.—This is one of the most remarkable turtles occurring within our limits, if not within the United States. It is rare in collections, and persons living along the lower Wabash ought to secure all the specimens possible. It may at all times be distinguished from the common Snapping-turtle by the three extra plates above those marginals which are placed just above the bridge. Its great head and its rapidly descending snout are also good marks. It is an exceedingly strong and fierce turtle, and a large one would be hard to manage. Mr. True states that he has known a specimen of perhaps forty pounds to bite the handle of a broom quite in two when enraged. They live principally on fish, but will no doubt devour almost any animal that may be so unfortunate as to come within range of their powerful jaws. One is mentioned (*4*, i, 415) as catching a bass about fourteen inches long and holding it down on a rock with his fore feet and greedily eating it. The breeding habits are not well understood. Agassiz figures the egg. It is spherical and an inch and three-eighths in diameter.

Family IX. KINOSTERNIDÆ.

Body rather narrow and high. The greatest height behind the middle, beyond which the outline descends rapidly. Marginal plates 23. Plastron moderately to well developed; formed of 8 bones, the entoplastron being absent. Plastral scutes 10 or 11; the gulars present and united or absent; the pectorals not meeting the marginals; abdominals cut off from marginals by two small plates on the bridge. Head large; jaws strong; snout projecting. Digits moderately developed and webbed. Five fingers and four toes with claws. Eggs elliptical.

Plastron narrow, its hinder lobe not more than one-half the width of the carapace. *Aromochelys*, p. 153.

Plastron wider, its hinder lobe considerably wider than one-half the carapace. *Kinosternon*, p. 154.

Genus AROMOCHELYS, Gray.

Aromochelys, Gray, 1855, *25*, 46; *Goniochelys* and *Ozotheca*, Agassiz, 1857, *4*, i, 423, 424; *Cinosternum*, in part, Boulenger, 1889, *84*, 33.

Shell of the young with a prominent keel, which may persist in the adult or more or less disappear. Plastron lacking much of filling up the opening of the carapace, the hinder lobe not more than one-half the width of the carapace. The lobes little movable on the middle portion, and the whole length of the plastron considerably less than that of the carapace. Suture between the pectorals longer than that between the humerals.

Gular scute present; head with yellow streaks from snout.
A. odorata, p. 153.
No gular scute; head with dark spots, no streaks of yellow.
A. carinata. Appendix.

Aromochelys odorata, (Bosc.).

Musk Turtle.

Testudo odorata, Bosc., 1803, *69*, 189; *Sternothærus odoratus*, Holbrook, 1842, *54*, i, 133, pl. 22; *Aromochelys odorata*, Gray, 1855, *25*, 46; *Ozotheca odorata*, Agassiz, 1857, *4*, i, 425, and ii, pl. iv, figs. 1–6; *Cinosternum odoratum*, Boulenger, 1889, *84*, 37.

Body of the young broadly oval and with a prominent keel, toward which the shell slopes roof-like. As the animal grows older the shell becomes proportionally narrower, the middle of the back more rounded, and the keel almost disappears. The first vertebral scute is long and narrow. Plastron narrow, leaving wide gaps between it and the carapace. The anterior lobe slightly movable on the transverse hinge. It extends forward from this hinge only about two-thirds the distance to the anterior end of the carapace. A small, triangular gular scute present. Suture between the humerals not quite equal to that between the pectorals. Posterior lobe not more than one-half the width of the carapace, falling considerably short of the hinder end of the shell; notched behind. Plastral scutes of the adults separated by tracts of soft skin.

Head large, snout projecting, jaws strong, the lower hooked. Toes extensively webbed. Soft skin everywhere provided with prominent fleshy papillæ. Tail of males projecting beyond the carapace, coiled at the tip, and furnished with a small nail. Posterior borders of the thighs and lower leg with each a patch of horny, sharp-edged scales.

Color of the carapace brownish or horn-color; uniform or spotted and striped with dark brown. Upper surface of head, neck and limbs brown; the inferior surfaces paler. Head with two yellow stripes running back from the snout; one over, the other below, the eye. Plastron yellow. A specimen that was taken in May at Lake Maxinkuckee had all the soft skin suffused with red.

Length of carapace 4 or 5 inches.

Distributed from Maine to Florida and west to Louisiana and Western Missouri. It is no doubt to be found throughout the whole State of Indiana. It is reported from Brookville (Hughes), Mt. Carmel, Ill. (State coll.), lakes of Northern Indiana (Dr. G. M. Levett), Lake Maxinkuckee and Marion county (Hay).

HABITS.—This is to be regarded as essentially an aquatic tortoise. It appears to be disposed to frequent the deeper parts of ponds and small lakes, since in such places it is safer than about shores or on the land. Its disposition is timid, and it prefers to seek safety in concealment or in retreat, to defending itself actively. Nevertheless its disposition is not altogether mild, since it will, when prevented from escaping, put out its head slowly and close its jaws on its assailant with a sudden snap. Holbrook states that it will bite severely, if provoked. In their native haunts they are often seen basking in the sun on some projecting rock or on some fallen tree, from which on the slightest alarm they drop off into the water. Their food is probably mostly or altogether of animal origin. They lay their eggs on shores in holes that they have dug in the sand with their hind feet. The eggs are from three to five in number, of an elongated elliptical shape, a little more than an inch long, and have a hard, smooth shell. One kept in confinement by Agassiz laid after the middle of June.

It takes its specific name from the strong, musky odor which it emits in life.

Genus KINOSTERNON, Spix.

Kinosternon, Spix, 1824, *90*, 17; *Kinosternum*, LeConte, 1854, *1*, 180; *Cinosternum* and *Thyrosternum*, Agassiz, 1857, *4*, i, 426; *Cinosternum*, Boulenger, 1889, *84*, 33.

Shell with the median keel indistinct even in the young; almost or entirely disappearing in the adults. Plastron with its anterior and posterior lobes movable on the middle fixed portion. Length of plastron almost equal to the length of the carapace. The width of the posterior lobe more than one-half the width of the carapace. Suture between the pectorals considerably shorter than that between the humerals.

Kinosternon pennsylvanicum, (Gmelin).

Eastern Mud Turtle.

Testudo pennsylvanica, Gmelin, 1789, *64*, ed. 13, 1042; *Kinosternon pennsylvanicum*, Holbrook, 1842, *54*, i, 127, pl. 21; *Thyrosternum pennsylvanicum*, Agassiz, 1857, *4*, i, 428, pl. iv, figs. 7-12 (y'g), and pl. v, figs. 16, 17; *Cinosternum pennsylvanicum*, Boulenger, 1889, *84*, 39.

Body oval, not much elevated. The young with three indistinct keels, a median and two lateral; these almost wholly disappearing in the adults; often a depression along the middle of the back. First and second vertebral scutes considerably longer than broad. Plastron well developed, lacking but little of filling up the opening of the carapace. The anterior lobe rounded, extending even beyond the end of the carapace, freely movable on the transverse hinge. Gular plate single, small and triangular. The humerals large, and the suture between them much longer than that between the pectorals. Hinder lobe of plastron movable on the fixed portion; notched behind. Head of moderate size; snout not greatly projecting; lower jaw hooked. Soft skin with a few or no fleshy papillæ; a pair of barbels behind the symphysis of the mandibles. Males with two patches of sharp-edged scales, one above the hollow of the knee, the other below. Tail of the males projecting beyond the shell, ending in a horny curved nail.

Color of the carapace horn-color or brown, with the sutures black. Plastron yellow or brown, with the lines of growth usually very distinct. Soft skin above brownish, with spots of yellow. A yellow stripe from the snout over the eye and back on the neck. Another from the eye to the corner of the mouth and to the angle of the jaw. Skin of the lower surfaces yellow. Size small, becoming probably not more than 5 inches from front to back of carapace.

Distributed from Canada to Florida and west to Texas and Western Kansas. In Indiana it has been taken by Mr. Robert Ridgeway, at Wheatland, in Knox county. Prof. W. S. Blatchley reports to me that he has collected a specimen at Terre Haute. Mr. Ridgway states that it is common on the borders of Monteur's Pond, in Knox county.

HABITS.—This tortoise is quite similar to the musk turtle in its mode of life. It remains about ponds and muddy ditches, where it can bury itself in the mud whenever it becomes alarmed. They are more inclined to passively withdrawing into their shells when attacked than is the musk turtle, since the shell forms a more perfect protection and their jaws are not fitted for the infliction of severe wounds. Their food consists of fish, insects and similar small animals. They are said to take the hook readily, but they nibble the angler's bait so slyly that their presence is not observed for some time. Their eggs are similar to those of *A. odorata*, but rather larger. They are laid in similar situations.

Family X. TESTUDINIDÆ.

Shell completely ossified, varying in form from broad to narrow and from high to depressed. Greatest height about the middle, whence the shell slopes in all directions, flaring at the borders. Carapace with the typical number of scutes, 4, 5, 4, with 25 marginals. Plastron large, composed of 9 bones, the entoplastron being present. Plastral scutes 11 or 12; the pectorals coming into contact with the marginals.

KEY TO THE GENERA OF *Testudinidæ*.

A. Parts of the plastron immovably sutured to one another and to the carapace.
    a. Alveolar surface of the upper jaw with a median ridge parallel with the cutting edge of the jaw.
                                                     Chrysemys, p. 156.
    aa. Alveolar surface without a ridge.
        b. Alveolar surface broad. Choanæ even with posterior borders of the orbits.   *Malaclemys*, p. 164.
        bb. Alveolar surfaces narrow. Choanæ between the eyes.
                                                       Clemmys, p. 168.

AA. Plastron with at least its anterior lobe movable on a transverse ligamentous hinge.
        c. Upper jaw not hooked. Shell two-thirds as wide as long.   *Emydoidea*, p. 170.
        cc. Upper jaw hooked. Shell three-fourths as wide as long.   *Cistuda*, p. 171.

Genus CHRYSEMYS, Gray.

Chrysemys, Gray, 1844, *91*, 27; Boulenger, 1889, *84*, 69; *Pseudemys*, Gray, 1855, *25*, 33; *Ptychemys*, *Trachemys*, and *Chrysemys*, Agassiz, 1857, *4*, i, 431, 434 and 438.

Shell moderately depressed. Bridge wide; the axillary and inguinal processes of the carapace (seen in the skeleton) well developed, the latter united to the 5th costal plate. Entoplastron lying wholly in front of the suture between the humerals and the pectoral scutes. Alveolar surfaces of the jaws broad, that of the upper with a median ridge parallel with the cutting edge of the jaw. Choanæ well toward the anterior border of the eyes. Hind legs stout, all the digits webbed beyond the bases of the claws. Skull with a bony arch bounding the eyes behind (temporal arch).

KEY TO THE INDIANA SPECIES OF *Chrysemys*.

A. Alveolar surface of the upper jaw broad, with the median ridge tuberculated. Upper jaw slightly or not at all notched in front. (*Ptychemys*, Ag.)
    a. Cutting edges of both jaws smooth or nearly so; tubercles of alveolar surface not prominent. Shell quite flat, deeply serrated behind. Plastron with its hinder border distinctly notched. Head small.
                                                      hieroglyphica, p. 157.

aa. Cutting edge of the upper jaw smooth, of the lower serrate. Median ridge of the alveolar surface coarsely tuberculated. Shell serrate behind. Head of moderate size. $\begin{cases} labyrinthica, \text{ p. } 158. \\ concinna, \quad \text{ p. } 159. \end{cases}$

AA. Alveolar surface of upper jaw of moderate width; the alveolar ridge not tuberculated, at most slightly denticulated. Upper jaw with a median notch, but no lateral teeth. (*Trachemys*, Ag.)

    a. Head rather small; posterior border of the shell very slightly serrated; plastron with a shallow notch behind. Plastral scutes each with a central blotch and a dark border. *troostii*, p. 160.

  aa. Head of moderate size. Shell serrated behind. Plastron with a distinct notch behind. Plastral scutes with a central blotch, their margins not bordered with black. *elegans*, p. 161.

AAA. Alveolar surface of the upper jaw rather narrow, widest behind; the median ridge not prominent. Upper jaw with a notch in front, on each side of which there is a small tooth. (*Chrysemys*.)

    a. Costal scutes alternating with the vertebrals, the transverse rows not straight.
      b. Costal scutes without red or yellow bands across them. *marginata*, p. 163.
      bb. Costal scutes crossed by red or yellow bands. *bellii*. Appendix.
  aa. Costal scutes placed even with the vertebrals, the rows across the carapace being straight. *picta*. Appendix.

Chrysemys hieroglyphica, (Holbrook).

Hieroglyphic Terrapin.

Emys hieroglyphica, Holbrook, 1842, *54*, i, 111, pl. 17; *Ptychemys hieroglyphica*, Agassiz, 1857, *4*, 1, 434; *Chrysemys hieroglyphica*, Boulenger, 1889, *84*, 76.

Head unusually small; snout somewhat projecting; upper jaw slightly notched in front; both upper and lower jaws smooth or slightly denticulated. Shell greatly depressed, and in large specimens without trace of keel. In specimens five inches long there is a slight keel. Shell sometimes smooth, occasionally longitudinally wrinkled. At its border, especially behind the thighs, the shell flares outward excessively, in some cases producing an actual concavity in the shell above. Hinder margin, deeply serrated. The bridge is narrow from front to back, the width being contained in the length of the plastron about three times or more. It rises little toward the carapace, and this contributes to the apparent flatness of the shell. Hinder border of the plastron with a deep notch. Longest suture that between the abdominals; the shortest, that between the humerals. Digits all strongly webbed. Hind feet very large and flat.

The ground color of the carapace varies from olive to dark brown. This is variegated with numerous lines and stripes of yellow. On the vertebrals the lines tend to run longitudinally. On the costals broad yellow bands divide each scute into three or four areas, inside of each of which are narrow concentric lines of the same color. The marginals are marked with yellow and brown. The plastron is yellow, with some splotches of brown on the bridge. The head, neck, feet and tail are all dark green, with numerous longitudinal bands of yellow. The length of the shell of large specimens is 12 inches.

Habitat from Georgia to Texas and north to the Wabash Valley.

Two shells of this species are in the State collection, which were sent from Mt. Carmel, Illinois, on the Wabash River. No doubt it will be found along the whole lower course of the Wabash. In the "Report of the State Geologist of Indiana" for 1875, page 499, Dr. G. M. Levette reports the occurrence of this species in the Kankakee River. Dr. Levette had given considerable attention to the study of our tortoises, and it is quite probable that he was correct in his determination of the species. I have had the opportunity of studying a number of specimens of this species in the National Museum.

Nothing appears to be known concerning the special habits of this terrapin. It is undoubtedly entirely aquatic, as are its immediate relatives.

Chrysemys labyrinthica, (LeS.).

Emys labyrinthica, LeSueur, MSS. in *113*, 13; *Malacoclemmys geographica*, in part, Agassiz, 1857, *4*, i, 436; Boulenger, 1889, *84*, 90; *Ptychemys labyrinthica*, G. Baur, MSS.

The type of this species was taken by LeSueur in the Wabash River, probably at New Harmony, and is now in the Museum d' Histoire Naturelle, at Paris. Both Agassiz and Boulenger regarded it as belonging to *Malaclemys geographica*, but it is evidently not this species. Duméril, in his description, states that the lower jaw is denticulated and furnished with a hook which fits into a corresponding depression in the upper jaw. He correctly compares the species with *C. hieroglyphica*, but says that it differs from the latter in the less elongated oval of the carapace and the elevation of the vertebral line, the shell of *hieroglyphica* being much depressed. The species received its name, as said by Agassiz, from the numerous meandering lines upon the bridge of the sternum. Not having seen specimens, I am unable to state how it differs from *C. concinna*. Dr. G. Baur, of Chicago University, to whom I am indebted for notes regarding it, states that the skull is much different from that of all other species. He believes that Prof. H. Garman's specimen from the Wabash River, described as *concinna*, belongs to LeSueur's *labyrinthica*. The two species are closely related, and specimens should be carefully sought along the Wabash and preserved.

Chrysemys concinna, (LeConte).

The Neat Terrapin.

Testudo concinna, LeConte, 1820, *62*, iii, 100; *Emys concinna*, Holbrook, 1842, *54*, i, pl. 8; *Pseudemys concinna*, Gray, 1855, *25*, 34; *Ptychemys concinna*, Agassiz, 1857, *4*, i, 432, pl. ii, figs. 4–6; *Chrysemys concinna*, Boulenger, 1889, *84*, 83.

Form of the shell somewhat variable in specimens of all ages, some having the greatest breadth at the middle, others at the hinder part; some are depressed, others more elevated. The young have a distinct keel, which is lost in half-grown specimens. The posterior border of the carapace is slightly serrated, the notches being between the scutes. Plastron with its posterior border with a distinct emargination; the hinder lobe not two-thirds the width of the carapace. Bridge wide, rising with moderate rapidity toward the carapace. Head of moderate size, the snout short and blunt. Upper jaw not at all notched in front; the cutting edge smooth; the alveolar ridge strongly tuberculated. Lower jaw with its sheath flat and rough on the outside, the cutting edge coarsely serrated, the tip with a sharp upturned point. Limbs well developed; all the digits webbed beyond the bases of the claws. Claws of the fore limbs of the males very long.

Color of the upper shell olive or brownish, with markings of yellow and dark brown. A yellow band usually runs down the middle of each costal scute. This usually sends off lateral anastomosing branches, which divide off the surface into a few large areas. Within these the yellow and brown are arranged in concentric lines. Both the upper and the lower surfaces of the marginals have eye-like spots of brown and yellow, one located across each suture. Across each scute there usually runs a yellow vertical band. Plastron almost uniform yellow, a few small spots of dusky on the anterior end, and about two on each bridge. Head, neck, legs, and tail brown, with many longitudinal stripes of yellow, or sometimes red. On the head there is a median stripe from the snout to the back of the head; another starting over the eye, widening on the back of the head; a stripe starting at hinder corner of the eye; another originating under the eye; and all, except the median stripe, running back on the neck. The lower stripe is met behind the corner of the mouth by a stripe from the middle of the lower jaw. At the tip of the jaw a stripe begins which further back divides into two, these including another yellow stripe.

The length of the shell may become as much as 16 inches.

The species ranges from North Carolina to Texas and north to Southern Indiana. Prof. Harry Garman, of Lexington, Ky., states (*61*, 1892, 185), that he received a fine, large specimen of this species from Dr. J. Schneck, of Mt. Carmel, Ills.; and he further says that several others have been observed in the same locality. The species will, therefore, be found along the lower part of the Wabash River, and possibly further north.

This species may always be readily distinguished from all others by the smooth edge of the upper jaw and the serrated edge of the lower.

HABITS.—Not much appears to be known about the habits of this terrapin. It is quite common in the waters of the more southern States. Mr. Fred. W. True, of the National Museum, states (*52* i, 155), that it seems to prefer brackish waters. Their diet consists principally of animal matter, and they are reported, in the South, to feed on certain species of worms, which they capture by inserting their claws into the worm-holes in the clay. This seems extremely doubtful. Agassiz found twelve eggs in the oviducts of one specimen. The eggs are of an elliptical form, about an inch and a half long and an inch in the shortest diameter.

Chrysemys troostii, (Holb.).

Emys troostii, Holbrook, 1842, *54*, i, 123, pl. 20; *Trachemys troostii*, Agassiz, 1857, *4*, i, 435; *Pseudemys troostii*, Cope, 1875, *12*, 53; *Chrysemys troostii*, Boulenger, 1889, *84*, 76.

Shell only moderately depressed; said by Holbrook to be "greatly depressed." There is only a trace of the keel in the adults. Behind the bridge the shell flares outward, but not so much as in *C. hieroglyphica*. The posterior border is only slightly serrated. The upper surface is, in adult specimens, somewhat wrinkled. The plastron has a broad shallow notch behind. The bridge is wide, but does not rise much toward the carapace. The longest suture of the plastral scutes is that between the abdominals, the shortest that between the humerals. The head is relatively small, flat above, and pointed. The cutting edge of the upper jaw is convex on each side, with a slight nick in front; the alveolar ridge is low and smooth. Lower jaw ending in a turned up tip. Fore and hind limbs well developed; the digits all completely webbed; the claws of the fore foot of the males very long and curved. The tail of the males very long.

The ground color of this species may be regarded as greenish horn-color above, yellow below. The scutes of both the carapace and the plastron are bordered with dark brown. On the carapace, within the areas thus formed, there may be a little black in splotches. Or this may increase in amount until nearly the whole scute is covered. This is especially true on the hinder half of the shell. On the plastron, besides the dark margin, each scute may have a central spot of black, and this by expanding may occupy most of the surface. This is also more likely to be the case on the hinder end of the plastron. On each of the two anterior scutes there is an eyelike spot, consisting of a circle of black inclosing another of the same color. The yellow of the plastron is to a considerable extent replaced by red. The soft skin of the head and upper side of neck is olive or dusky, varied with numerous fine anastomosing lines of pale yellow. At the back of the eye a stripe begins and runs back on the neck. This stripe is bright red, not well defined along the edges, but seeming to run into the surrounding dark color. In some cases the whole of the back of the head is red. The feet, legs, tail and lower side of the neck are ornamented with broad yellow or green stripes.

The length of the shell may reach 9 inches, probably more.

Mississippi River and its tributaries from the Gulf to Northern Missouri. It has been sent to the National Museum from Wheatland, Ind., by Mr. Robert Ridgway, to whom we are indebted for knowledge of many rare reptiles of this State.

This is a very beautiful and a characteristically marked species. It may readily be distinguished from *C. elegans* by the brown borders of all the scutes, and the absence of yellow stripes on the carapace. Both have a blood-red stripe along the neck.

HABITS.—Little is known, beyond the fact that it is aquatic. It presents a good subject for study to herpetologists who live on the lower reaches of the Wabash.

Chrysemys elegans, (Wied).

Elegant Terrapin.

Emys elegans, Wied, 1839, *63*, i, 213; *Trachemys elegans*, Agassiz, 1857, *4*, i, 435, pl. iii, figs. 9–11; *Pseudemys elegans*, Cope, 1875, *12*, 53; *Chrysemys scripta*, var. *elegans*, Boulenger, 1889, *84*, 78.

Shell broad and depressed, the young with a moderate keel, which disappears in the adults. Carapace serrated behind; a slight emargination in each scute and deeper ones between them. Surface of the costal scutes smooth or sometimes slightly wrinkled longitudinally. Nuchal scute very narrow. Plastron with its posterior border with a broad shallow notch; the width of the hinder lobe being hardly two-thirds the width of the carapace. Bridge wide, rising rapidly to the margin of the carapace. Longest suture of the plastron that between the abdominals, the snortest that between the humerals. Head of moderate size; snout short and rather blunt. Edge of upper jaws convex along the sides, notched in front; smooth; the alveolar surface with a low smooth ridge. Lower jaw smooth or slightly denticulated. The tip curved upward. Limbs well developed; all extensively webbed; claws of forefoot of males very long and somewhat curved. Tail of moderate length.

Color of the carapace olive, with lines and spots of yellow and black. On the vertebral scutes the lines run mostly lengthwise, on the costals transversely. Down the middle of each costal scute runs a yellow band of varying width. Parallel with it are other lines and bands of black and yellow, narrow or wide. On both the upper and the lower surfaces of the marginal scutes are sutural spots consisting of concentric circles of yellow and black. Between them a yellow band crosses each marginal. The plastron is yellow, with a black blotch on each scute, these often ocellated with yellow. The spots on the bridge usually confluent. Head with numerous narrow stripes of greenish or yellow. A broad stripe starts under the eye and runs back on the neck, being met at the angle of the jaw by a stripe from the middle of the lower jaw. Another stripe, often blood-red, starts at the posterior corner of the eye and runs back on the neck. The stripe is wanting which in *C. concinna* starts above the eye and extends on the neck. The legs and tail are striped with yellow.

Length of the shell in large specimens about 10 inches.

This species has been found inhabiting the territory from South Carolina to Mexico, and north along the tributaries of the Mississippi to the Yellowstone. In Indiana it has been taken at New Harmony (Sampson's coll.); in the Wabash River at Mt. Carmel, Ills. (L. M. Turner). The species was originally described by Max Von Wied from specimens taken near New Harmony. About July 1, 1892, I took a specimen in the Tippecanoe River at Winamac.

Dr. Boulenger, as above cited, regards this terrapin as only a variety of *C. scripta* (*Trachemys scabra*, Ag.). However, at present it appears to me that there are sufficient differences in both the young and the adults of the two forms to justify their being regarded as distinct species. Their geographical distribution is likewise different. The specimen of *C. elegans* reported by Dr. Yarrow (*10*, 33) from Oakley, S. C., is a young *C. scripta*.

HABITS.—Quite as little is known about the manners of life of this species as of most of the other aquatic turtles. Agassiz figures the egg. It is elliptical in form, an inch and a half in its long and seven-eights in its short diameter. This naturalist has also observed that this turtle has a voice, as he believed most turtles have. It is said to "emit a piping note." Dr. J. Schneck, of Mt. Carmel, Ills., kept a young *elegans* for more than two years, during which time it partook freely of food, but made no perceptible growth. Prof. F. W. Craigin (*49*, 1, 101) thinks that these and other turtles are sometimes killed by minks and other carnivorous animals. To the attacks of such enemies may be due the great timidity of turtles, which seem to have so few enemies.

Chrysemys marginata, Ag.

Western Painted Tortoise.

Chrysemys marginata, Agassiz, 1857, *4*, i, 439, pl. i, fig. 6 and pl. v, figs. 1–4; *Chrysemys cinerea*, Boulenger, 1889, *84*, 73.

Shell broad and depressed, broadest behind the middle; the shell flaring considerably posteriorly; its surface very smooth; no traces of a keel, even in the young. Scutes of the carapace arranged as usual among tortoises, the suture between the costals meeting the lateral border of the proper vertebral about its middle. Vertebrals 2 and 3 wider than long, but narrower than the costals. Anterior border of the carapace often with a few dentations; the posterior border not serrated. Plastron broad and flat, truncated before and behind; the anterior end often denticulated. Bridge wide, flat, and rising rather rapidly to the margin of the carapace. Head of moderate size; snout not much projecting. Jaws with smooth cutting edge, the front with an evident notch, on each side of which is a small tooth. Alveolar ridge feeble. Lower jaw little upturned. Limbs with moderate development; the digits webbed to the claws. Tail of moderate length, that of the males longest.

The color of the carapace is usually dark green. The hinder border of the costal and vertebral scutes is narrowly bordered with black. On the anterior border of the same scutes, and lying immediately against the black margin, are slightly wider lines of bright red (yellow in alcoholic specimens). These red or yellow lines do not join so as to form straight lines across the back. A very narrow line of red runs along the middle of the back. Upper surfaces of the marginal plates with many crescent-shaped marks. of bright red. Lower surfaces of the marginals black, with large splotches of blood-red and bright yellow. Plastron bright yellow or brownish red, with a large dusky blotch occupying its central portion. Soft skin of head, legs and tail dark olive, with red stripes. On the occipital region are two large waxy yellow spots, nearly as large as the eye; these prolonged backward into two narrower pale yellow stripes. Another short yellow stripe from the upper corner of the eye; another from the lower side of the eye and running back on the neck. Two red stripes on the front of the fore legs, and similar ones on the posterior surfaces of the thighs. Besides these, there are numerous small spots of red all over the soft parts. All the red fades to yellow in alcohol.

The usual length of the shell is about 4 or 5 inches; a length of 7 inches may be attained. This species is an inhabitant of the Northern

States of the Mississippi Valley from Ohio to Kansas, and north to Lake Superior. In Indiana it is to be found everywhere. I have proofs of its occurrence at so many points that it seems unnecessary to state them.

HABITS.—This is at once our most beautiful and most common species of tortoise. It is, however, probably less well known than the Snapping-turtle, because of its strictly aquatic mode of life and its excessive timidity. It appears to prefer to abide in ponds, pools, and the sluggish parts of our streams. In such places it may be often seen lying with its fellows on some fallen tree-trunk or on some projecting stone, basking in the sunshine. The senses of sight and hearing appear to be acute, for it easily takes alarm and tumbles into the water and disappears. It is then often to be found buried in the mud close to where it entered the water. It is an entirely harmless turtle, and can hardly be provoked to bite, and its effort is then a feeble one. The food of the Western Painted Turtle probably consists of insects, tadpoles and other feeble and small animals.

Smith (*18*, 665) states that in Michigan this turtle has been found out of its winter quarters as late as October 22, and in the spring on March 31. From tortoises that have been marked, it appears that all these animals wander but short distances from where they have been hatched. According to Agassiz' figures, the eggs of this tortoise are about an inch and a quarter long and nearly seven-eighths in the shorter diameter. Many interesting things are to be found concerning the closely related *C. picta* in Agassiz' work on the Testudinata of North America (*4*).

Genus MALACLEMYS, Gray.

Malaclemys, Gray, 1844, *91*, 28; 1855, *25*, 37; *Graptemys* and *Malacoclemmys*, Agassiz, 1857, *4*, i, 436, 437; Boulenger, 1889, *84*, 88.

Shell depressed, with a distinct keel. Bridge wide, with the axillary and inguinal processes well developed, the latter united to the 5th costal plate. Entoplastron lying wholly in front of the suture between the humerals and pectorals. Jaws with the alveolar surface broad to very broad and entirely without a median ridge. Choanæ behind the level of the eyes. Skull without a bony temporal arch. Digits extensively webbed.

Keel strongly tuberculated; an elongated, transverse, yellow streak behind each eye. Carapace strongly serrated behind.

pseudo-geographica, p. 165.

Keel not tuberculated; a triangular yellow spot behind each eye. Carapace feebly serrated behind. *geographica*, p. 166.

Malaclemys pseudo-geographica, (LeS.).

LeSueur's Map Tortoise.

E. pseudogeographica, LeSueur, MSS. in Holbrook, 1842, *54*, i, 103, pl. 15; *Graptemys lesueurii*, Agassiz, 1857, *4*, i, 436, pl. 2, figs. 10-12; *Malacoclemmys lesueurii*, True, 1875, *10*, 34; Boulenger, 1889, *84*, 91 (not *Emys lesueurii* of Gray).

Shell oval, depressed, rising roof-like to the distinct medial keel. Posterior border of some or all of the vertebral scutes with each a prominent tubercle, largest on the 2d and 3d vertebrals. Shell strongly serrated behind. Nuchal with a notch in its hinder border. Plastron with its hinder lobe not much over one-half the width of the carapace; a broad shallow notch in its hinder border. Bridge broad and flat, rising little toward the carapace. Head of the males small; that of the females rather large. Snout not at all projecting. Cutting edge of the upper jaw smooth, convex, the jaw not notched in front; the alveolar surface of moderate width, wholly separated in front by soft skin. Lower jaw with smooth, concave cutting edge, not hooked at the tip. Limbs well developed, the digits webbed to the bases of the claws. Tails of the males, as with most turtles, bringing the vent beyond the edge of the carapace.

The color of the upper surface of the carapace is olive or occasionally brownish. Usually there are no black spots on the carapace; but occasionally there is a blotch, as if made with the inked end of the finger, on each of the larger scutes and on most of the marginals. Over all the scales there is a network of greenish lines, often obscure, which divide each scute into about 4 or 5 areas. The plastron is yellow, with a little clouding or mottling of brown, or with many irregular and obscure stripes and lines of dark color. Bridge almost uniform brown or with numerous streaks of yellow and brown. Head, neck, limbs, and tail dark green, with many stripes of yellow, and many rows of small yellow spots. Behind the eye there is a very characteristic transverse streak of yellow, sometimes short, sometimes turned forward beneath the eye. When the streak is short, there is a yellow dot under the eye.

The shell of this species may reach a length of 10 inches. Adult female specimens will average 8 inches or less; males usually smaller.

This is a species belonging to the Mississippi Valley, ranging from Ohio to Kansas, south to Louisiana, north to Wisconsin. It doubtless occurs throughout the State of Indiana; nevertheless, I did not find it at Lake Maxinkuckee, and Dr. Levette does not give it in his list of turtles found in the northern part of the State. It is abundant about New Harmony (LeSueur, Max. V. Wied, and Sampson's coll.); found at Brookville (Hughes); Terre Haute, rare (Blatchley); Monroe county (Bollman).

This species is quite distinct from *M. geographica*, as shown by the much narrower alveolar surfaces of the jaws, the strongly tuberculated vertebral scutes, the more distinct keel, the transverse streak behind the eye, the coarser network of lines on the carapace, and the greater amount of brown on the plastron. Some authors, among them recently Dr. Boulenger and Prof. Harry Garman (*42*, xxii, 70) have given as distinctive characters the large head of *geographica* and the smaller head of *pseudo-geographica*. The size of the head is a sexual character in both species, the males having a small head, the females a large head. Females of the two species of the same size have heads of approximately the same width, and the same is true of the males. Of this I have satisfied myself by measurement of specimens in the National Museum and in my own collection. The head is, however, variable in relative size in different individuals of the same sex. Moreover, it will be found, I think, that the males average considerably smaller in size than do the females.

HABITS. — This is an eminently aquatic tortoise, spending its life in rivers, lakes and ponds, and coming out of the water only to bask in the sun on some rock or fallen tree, or to deposit its eggs. The food of LeSueur's Map-Turtle has hitherto been regarded as animal in nature, such as small fishes, reptiles and the like, but Prof. Garman states that the digestive canal of most of the specimens that he observed were filled with the bulbs of a sedge. In some cases, however, it was found to have eaten crayfishes. The eggs are large, being an inch and a half in the longest and an inch in the shortest diameter. According to Agassiz this species deposits its eggs earlier in the season than any others of our turtles. At Natchez, Miss., one was found to have laid her eggs as early as the first of June. It may here be stated that Agassiz concluded that our fresh-water turtles do not lay eggs before the eleventh or fourteenth year.

This species does not appear to be employed to any considerable extent as food, yet there seems to be no reason why its flesh should not be as savory as that of many species which are highly esteemed.

Malaclemys geographica, (LeS.).

Geographic Terrapin; Map Tortoise.

Testudo geographica, LeSueur, 1817, *2*, 86, pl. v; *Emys geographica*, Holbrook, 1842, *54*, i, 99, pl. xiv; *Graptemys geographica*, Agassiz, 1857, *4*, i, 436, pl. ii, figs. 7–9; *Malacoclemmys geographicus*, Cope, 1875, *12*, 53; Boulenger, 1889, *84* 90; *Emys lesueurii*, Gray, 1831, *112*, 31.

Shell depressed, and keeled in small individuals, becoming more elevated, higher and more rounded in full grown adults. Keel with rudimentary tubercles, this evidence of the keel persisting even in adults.

Carapace feebly serrated posteriorly. Nuchal narrow, its hinder border notched. Carapace rounded behind in the females, more pointed in the males. Plastron with its posterior lobe about two-thirds the width of the carapace; distinctly notched behind. Bridge wide, rising little toward the carapace. Limbs well developed, scaly, the digits well provided with webs. None of the claws of the male much elongated. Head of the males small, that of the females large. Snout not at all projecting. Upper jaw with the cutting edge smooth, somewhat sinuated, not notched in front; the alveolar surface very broad, united with its fellow back nearly to the choanæ. Lower jaw flat, its alveolar surface resembling that of the upper jaw. The jaw not hooked at the tip.

Ground-color of the carapace dark olive. Over all the scutes there is a network of greenish lines, so that each of the large scutes is divided into about 8 to 10 areas. Under side of the marginals with sutural spots of dark green, which enclose irregular lines of yellow. Head, neck, limbs and tail dark green, almost black, with numerous lines and streaks of greenish yellow. Behind the eye is a somewhat triangular spot of greenish yellow, often elongated backward. Plastron yellow, with the sutures of the scutes marked with a dark line. Occupying the center of the plastron is a large lyriform blotch of brown, which looks as if the color had almost faded out.

The size of this species averages larger than that of *M. pseudo-geographica*, full grown specimens being about 10 inches long. It may become still larger.

This species is distributed from Pennsylvania and New York to Michigan and Arkansas. It is found no doubt in all the streams and lakes of Indiana. Known localities are New Harmony; Brookville; North Manchester, Wabash county (A. B. Ulrey); Lake Maxinkuckee; Eel and St. Joe rivers; Terre Haute (Blatchley); Fall Creek, Marion county (W. P. Hay); Kankakee River at English Lake; Tippecanoe River at Winemac.

This turtle can be readily distinguished from any other species occurring in Indiana by the extremely expanded masticatory surface of the jaws. From *M. pseudo-geographica* it may be distinguished by the reduced keel, especially of the large females, the rudimentary tubercles of the keel, and especially by the triangular spot behind the eye. As stated in the description of *M. pseudo-geographica*, it is the head of the female that is large; that of the male is little, if any, larger than males of the same size of *pseudo-geographica*. My observation is that the females are usually much larger than are the males. Dissections made of 7 specimens taken at Lake Maxinkuckee showed that 4 were females, all with carapace more than 6 inches long; the others were not over 4 inches long, and all were males. I think that males may become somewhat larger than these, but not nearly so large as the largest females.

HABITS.—The mode of life of the Map Turtle is as thoroughly aquatic as that of its relative, *M. pseudogeographica*. It probably never, unless compelled to do so, leaves the immediate vicinity of its native stream. Holbrook states that it is bolder and more active than most other turtles, those that he had seen approaching even the snapping turtles in their disposition to bite when disturbed. The food of this species consists of animals of various kinds. Prof. Harry Garman (*42*, xxii, 70) states that an examination of the contents of the alimentary canal showed that the food consisted exclusively of mollusks, the young eating the thinner shelled species, the adults the larger and thicker shelled kinds. At Lake Maxinkuckee three persons caught about 30 specimens of this species in a few hours. Without probably an exception they were found near the shores, where there were great numbers of the water-breathing univalves. After a number had been kept for a few days in a tub there were found in it large numbers of the opercles of such mollusks; and in the intestines of one were the remains of a crayfish, some fish scales, and what appeared to be the cases of some kind of caddis-worm. Its broad masticatory surfaces are well fitted for crushing the shells of mollusks.

The eggs of this species, as figured by Agassiz, appear to be somewhat smaller than those of LeSueur's tortoises. I have found 16 eggs in a large female. DeKay states that the flesh of this tortoise is good for food. Where they are abundant they might be turned to good account.

Genus CLEMMYS, Wagler.

Clemmys, Wagler, 1830, *75*, 136; *Boulenger*, 1889, *84*, 100; *Nanemys, Calemys, Glyptemys*, etc.; Agassiz, 1857, *4*, i, 442 seq.

Shell moderately to strongly depressed. Bridge wide, with strong axillary and inguinal processes of the plastron just reaching the 1st and 5th costal plates. Entoplastron crossed by the suture between the humerals and the pectorals. Alveolar surfaces of the jaws narrow and without a median ridge. Choanæ toward the front of the eyes. Skull with a bony temporal arch. Digits more or less extensively webbed.

Clemmys guttata, (Schneider).

Speckled Tortoise.

Testudo guttata, Schneider, 1792, *120*, x, 264; *Emys guttata*, Holbrook, 1842, *54*, i, 81, pl. 11; *Nanemys guttata*, Agassiz, 1857, *4*, i, 442, pl. i, figs. 7–9; *Chelopus guttatus*, Cope, 1875, *12*, 53; *Clemmys guttata*, Boulenger, 1889, *84*, 109.

Shell oval, widest behind, rather depressed, no traces of keel in the adult, little trace even in the young. Nuchal scute very narrow. Plastron large; the hinder lobe about three-fourths the width of the carapace, with a shallow emarganation in the posterior border. The anterior lobe truncated, not movable on a transverse hinge. Bridge rather narrow, not more than half the width of the hinder lobe of the plastron, rising rather rapidly to the carapace. Plastron of the male concave. Head of moderate size, covered with a hard smooth skin. Snout not at all projecting. Upper jaw notched in front; the alveolar surface very narrow. Lower jaw with the sheath externally very wide; the tip upturned. Choanæ well forward, under the front of the eyes. Legs and feet all covered with scales, those of the front limbs large and overlapping. Feet not large, the claws rather short, the web not extensive. Tail long, that of the male bringing the vent beyond the carapace.

The general color of the carapace is black. Sometimes there appear to be patches of reddish brown showing through the darker. On each scute there appear from one to a dozen round spots of bright orange, each larger than the pupil. The plastron is red, orange and black, the black generally predominating. The orange usually occupies the center of the plastron and the margin. Head above black, with orange dots. Generally there is a large spot of orange just above the ear. The neck is black, with more or less red mingled therewith. The shoulders are extensively red or orange. The upper surfaces of the limbs are black, with dots of yellow and red; the lower surfaces red and orange. The tail is black, with red at the base. Length of shell 4 or 5 inches.

Distribution from New England to North Carolina, west to Indiana. In this State it appears to be found only in the northern portion among the numerous lakes, streams and swamps found there. Dr. G. M. Levette first found it in that region, reporting it as occurring "in ditches around Kendallville, and doubtless over the whole region." Two specimens were picked up one morning at Lake Maxinkuckee in May, 1891, by members of the Indiana Academy of Science. Taken also at Rochester, Fulton county, by Dr. Vernon Gould; English Lake (Dr. Baur).

HABITS.—This little turtle is less exclusively aquatic than any of those that have been described, except the snapping-turtle. It seems to delight in being in the neighborhood of swamps and sluggish streams, and it probably spends the greater part of its time in the water. Nevertheless it often leaves the water, and it may be picked up while it is making its journeys. It is a very harmless animal, and deserves protection. Holbrook says that it is timid and gentle, and can easily be domesticated. When at freedom they collect in numbers on objects above the water and enjoy the sunshine; but if any fancied enemy is seen approaching they slide off rapidly into the water and soon bury themselves in the mud. Their food is said to consist of tadpoles, young frogs and other weak animals. On land they devour earthworms, crickets and grasshoppers.

Their eggs are few in number; never, according to Agassiz, exceeding three or four. They are about an inch and a quarter long and three-quarters in the shorter diameter. The eggs are laid about the 20th of June, in a perpendicular hole dug by the use of the hind legs. After the eggs are deposited the dirt is pushed back over the opening so as to conceal it entirely.

Genus EMYDOIDEA, Gray.

Emys, Duméril, 1806, *119*, 76; Agassiz, 1857, *4*, i, 441; Boulenger, 1889, *84*, 114; *Emydoidea*, Gray, 1870, *25*, Sup. 19; Baur, 1890, *22*, xxiii, 1099.

Shell moderately elevated. Bridge narrow, the plastron not sutured to the carapace, but united to it by ligament; therefore, more or less movable on it. Plastron divided by a transverse hinge at front end of bridge into two lobes, which are movable on each other. Axillary and inguinal processes of the plastron short, the latter just reaching the fifth costal plate. Entoplastron reaching but hardly intersected by the suture between the humerals and the pectorals. Alveolar surface narrow, without a median ridge. Choanæ between the eyes. Skull with a bony temporal arch. Digits webbed.

According to Dr. Baur, this genus differs from the Old World *Emys* in that the frontals enter the orbits, and the rib-heads are long, as in *Chelydra*.

Emydoidea blandingii, (Holb.).

Blanding's Tortoise.

Cistuda blandingii, Holbrook, 1842, *54*, i, 39, pl. 3; *Emys meleagris*, Agassiz, 1857, *4*, i, 442, pl. iv, figs. 20–22; *Emys blandingii*, Boulenger, 1889, *84*, 114; *Emydoidea blandingii*, Gray, 1870, *25*, sup. 19; Baur, 1890, *22*, xxiii, 1099.

Shell elongated oval, widest just behind the middle; rather high, convex, and without keel. Carapace not serrated behind. Plastron large, entirely closing the shell; movable on carapace on the ligamentous hinges, the two lobes movable on each other on a transverse hinge covered by the suture between the pectorals and the abdominal scutes; the posterior lobe somewhat excavated behind. Posterior border of the entroplastron reaching the humero-pectoral suture, but not intersected by it. Bridge narrow and very short, almost obliterated   Head long and wide, the eyes opening somewhat upward. Upper jaw with the cutting edge convex at the sides and notched in front. The alveolar surface narrow. Lower jaw with narrow alveolar surface and with a hooked tip.

Limbs, including the feet, scaly; the toes short and provided with a narrow web. Tail covered with scales; that of the male about two and two-thirds times in the length of the shell, that of the female shorter.

Color of the shell above, dark green to black, each scute with several round, triangular or oblong spots of yellow or orange, those of the marginals largest; all, however, sometimes wanting. Plastron yellow, with the outer posterior portion occupied by a blotch of brown. This blotch may expand so as to take in almost the whole scute. Head and neck above and along the sides dusky, with numerous yellow dots; chin, throat and under side of the neck yellow. Legs yellow, with mottlings of brown. Tail striped longitudinally with yellow and brown.

The carapace may attain a length of nine inches, but this is uncommon.

This species is wholly Northern in its distribution, being found from Massachusetts and Canada to Northern Illinois. In Indiana it occurs only in the region of lakes in the northern portion. Dr. Levette (*93*, 1875, 499) reports it as occurring "sparingly in the northern parts of Lagrange and Steuben counties." A live one was seen at Lake Maxinkuckee in May, and I have a shell of one that was taken at Rochester, Fulton county, by Dr. Vernon Gould, of that place. It does not appear to be rare in that region. It is also common at English Lake, in Starke county.

HABITS.—Not much accurate information has been gathered concerning the habits of this tortoise. It is probably somewhat less aquatic than the speckled tortoise, *Clemmys guttata*, yet it undoubtedly prefers the neighborhood of streams and ponds. I find no account of its food, but this is probably of an animal nature. Its eggs, as figured by Agassiz, are large and oval, measuring an inch and three-eighths by almost an inch. There are from seven to nine of them laid together once a year. The shell is thick, smooth and hard. According to Agassiz's figures there are no yellow or orange dots on the shell of the very young. In this respect they are in contrast with the young of *Clemmys guttata*, which are said to have the spots developed long before leaving the egg, even before the lungs are developed.

This species is to be distinguished from the box-tortoise by the more elongated, less elevated and less convex shell, the posteriorly notched plastron, and absence of anything like a hook to the upper beak.

Genus CISTUDA, Fleming.

Cistuda, Fleming, 1822, *115*, ii, 270; *Cistudo*, Bonaparte, 1830, *116*, 162; Agassiz, 1857, *4*, i, 444; Boulenger, 1889, *84*, 115.

Shell high and very convex. Plastron united to the carapace by ligament and movable on it; the axillary and inguinal processes rudimentary. Plastron divided by a transverse hinge into two movable lobes, the hinge covered by the suture between the pectoral and abdominal scutes. Entoplastron cut by suture between the humerals and the pectorals. Alveolar surface of jaws narrow, without median ridge.

Upper jaw with the beak projecting downward, notched or not.   Choanæ
between the eyes.   Skull without a bony temporal arch.   Digits with
short or no web.
  Shell with traces of a keel, rounded above; no bridge.
                                                         carolina, p. 172.
  Shell without traces of a keel, flat above; a distinct bridge.
                                                         ornata. Appendix.

Cistuda carolina, (Linn.).

Box Tortoise.

Testudo carolina, Linnæus, 1758, *64*, x, 198; *Cistudo carolina*, Gray, 1831, *112*, 18; Holbrook, 1842, *54*, i, 31, pl. 2; Boulenger, 1889, *84*, 115; *Cistudo virginea*, Agassiz, 1857, *4*, i, 445, pl. iv, figs. 17-19.

Shell broadly oval, sometimes four-fifths as wide as long; high and very convex; extremely solid. On at least the posterior part of the carapace are evidences of a keel; this in the young quite distinct. Margin of the carapace sloping rapidly upward from transverse hinge of the plastron. Plastron large, tightly closing the opening of the carapace, consisting of two lobes movable on each other and the carapace. The bridge entirely abolished; no axillary or inguinal scutes. The plastron rounded in front and behind. Head of moderate size, the snout not projecting; upper jaw with the cutting edge drawn down in front into a hooked beak, the hook not notched; the alveolar surface narrow. The lower jaw turned upward at the tip. Limbs and feet scaly, especially the anterior. Claws stout; the web between the digits narrow. Tail short. Scutes sometimes very smooth, sometimes showing distinctly the concentric lines of growth.

The colors of the carapace are yellow and brown or black. Sometimes the darker color predominates, sometimes the yellow. Usually the ground is brown or reddish brown, while the yellow appears as spots of various shapes; often radiating from the point of growth of the scute. The ground color may appear to be yellow, relieved with black spots. The plastron is variously ornamented with black and yellow. The young have a single yellow spot on each of the scutes of the carapace. The head, neck, limbs, and tail are brown, with numerous spots of yellow and orange. Often the scales of the fore-legs are especially bright yellow.

The length of the carapace is about 4 or 6 inches in full grown examples.

This tortoise is distributed from New England to the Gulf and westward to Texas. It inhabits the whole of Indiana, and appears to be especially abundant in the southern portion. New Harmony (Sampson's coll.); Brookville (Hughes and Butler); Monroe county (Bollman); Terre Haute (Nor. Sch. coll.); Lafayette and Westfield (F. C. Test); Jefferson, Marshall, and Marion counties (Hay); Wabash county (C. Ridgley).

HABITS.—This species is a thoroughly terrestrial animal; so much so that the statement has been made that it never goes near the water, can not endure even rain. This is a mistake, however, as I have seen the tortoise in a small, shallow rivulet which it might easily have avoided. On the other hand, it appeared to be enjoying the bath. Mr. Ed. Hughes of Brookville tells me that he too has seen them in the water; but also that he has seen them dead in deeper water, as though they had drowned. Considering the great thickness and weight of their shells it is not to be thought that they can swim readily or even at all.

These animals are entirely harmless, and when disturbed, retire within the shell and submit passively to their captor. They may be regarded as comparatively feeble animals, and in their thick, strong shells, which may be almost hermetically closed, we see a due compensation for their indifferent powers of self-defense.

The food of the box-tortoise appears to be of a mixed nature. Holbrook states that it feeds on insects, such as crickets, etc.; but he mentions LeConte's statement that they feed on fungi, such as *Clavaria*. Mr. Ed. Hughes of Brookville says that he dissected one and found in its stomach what appeared to be vegetable matter, but no insects. Max. Von Wied states (*103*, xxii, 6) that they greatly love cucumbers and lettuce, and do great injury to these plants. They are said to be very fond of mushrooms. Holbrook further says that this tortoise may be easily domesticated, and will eat whatever is offered it, as bread, potatoes, apples, etc. The notion that it will destroy mice and serpents as food he properly regards as improbable. The eggs of this species are of the usual shape, oval, about an inch and a half by three-fourths of an inch. They number from four to six, have a rather thin shell, and are laid about the latter part of June or later. During the winter these tortoises, like all others in our climate, remain buried in the earth. They appear to have been favorites for persons who attempt to secure immortality of name by engraving their names on terrapins' backs. From this practice something has been learned of their longevity. Dr. J. Schneck of Mt. Carmel, Ills., states (*22*, 20, 897,) that one at Albion, Ills., had had some initials engraved on it in 1824. It was found in the same vicinity in 1865, and marked with an additional letter. Again in 1885 it was seen within a half mile of the spot where it was liberated 20 years before. All the markings were quite distinct. Other cases of the kind prove that this tortoise lives a long time, and furthermore that it does not wander far from its early home.

APPENDIX.

The following species seem to require description here on account of the fact that they have been found not far from the borders of Indiana, and may, therefore, yet be taken within our limits.

Ambystoma xiphias, (Cope).

Sword-tailed Salamander.

Amblystoma xiphias, Cope, 1866, *1*, 192; 1889, *51*, 87, with figures; Boulenger, 1882, *28*, 45.

The only known specimen of this species is in the National Museum, and was taken at Columbus, Ohio. It is a very close relative of *A. tigrinum*. The head is narrower, the width being contained in the distance from the snout to the groin 4.5 times. The lower jaw projects prominently beyond the snout. The tail is longer than in most specimens of *A. tigrinum*, being considerably longer than the rest of the animal. The color, also, is different, yellow predominating in *A. xiphias*, dusky in *A. tigrinum*. Any specimens of apparent *tigrinum* which have peculiarities approaching those here mentioned should be carefully preserved and examined. Prof. W. S. Blatchley has shown me a specimen from Terre Haute which has the color peculiarities of *xiphias*, but it lacks the projecting jaw and the very long tail. It appears to be a true *A. tigrinum*.

Ambystoma talpoideum, (Holbrook).

Mole Salamander.

Salamandra talpoidea, Holbrook, *54*, v, 73, pl. 24; *Amblystoma talpoideum*, Cope, 1889, *51*, 52, with figures; Boulenger, 1882, *28*, 40.

Most of the known specimens of this species are from the Southern States, but it has been sent to the National Museum by R. Kennicott, from about Cairo, Ill. It is, therefore, to be sought for in the southwestern portion of our own State.

It is the smallest, stoutest, and most clumsily constructed of the species of the genus. The head is described as being broad and large, the width being contained in the length to the groin 3.5 times. There are only ten costal grooves. The tail is very short, being contained in the rest of the length 1.5 times. The color is a light brown, paler below, with sprinklings and marblings of silvery or leaden gray. There are some obscue dark spots on the back and tail. The length of the animal when full grown is less than four inches. Prof. Cope states that it lives in damp places below logs and stones.

Plethodon æneus, Cope.

Bronzy Salamander.

Cope, 1881, *22*, 878; 1889, *51*, 143.

Form and proportions as in *P. glutinosus*. Vomero-palatine patches in two straight lines, which do not pass out behind and beyond the choanæ. Costal grooves 13. The color is black, but the upper surface is thickly spotted with large, yellowish-green blotches. These occupy almost the whole surface of the head. The spots on the sides are considerably smaller. The lower surface is dusted with yellow.

This species was described from a single specimen which was found by Prof. Cope at the mouth of Nickajack Cave, located at the point where the three States of Georgia, Alabama and Tennessee adjoin. Dr. L. Stejneger informs me that the species has more recently been found in Lee County, Virginia.

Genus GYRINOPHILUS, Cope.

Gyrinophilus, Cope, 1869, *1*, 108; 1889, *151*, 154.

Vomero-palatine teeth in two transverse series which pass out beyond the choanæ. Parasphenoidal teeth present, in two parallel bands, these meeting the vomero-palatines at a right angle. Tongue with a free edge all around and standing on a central stalk. The premaxillary bones not ankylosed. Digits 4-5.

Gyrinophilus porphyriticus, (Green).

Salmon Triton.

Salamandra porphyritica, Green, 1827, *102*, i, 3; Cope, 1889, *51*, 155; *Spelerpes porphyriticus*, Boulenger, 1882, *27*, 64.

The type of this genus is *G. porphyriticus* (Green). This is a native of the Alleghany Mountains from the Adirondacks to Georgia, but since it has been reported from Columbus, Ohio, it is possible that it may yet be found in Indiana. I therefore add a short description of it. For further details regarding it see Cope's "Batrachia of North America."

Body elongated, slender and depressed; there is little constriction at the neck, and the tail tapers gradually to tip. The head is flat, and the upper jaw projects beyond the lower. There is a prominent ridge running from the eye to the edge of the jaw outside the nostril. The eyes are large and prominent. There are sixteen costal grooves, not counting one immediately in the axilla. There is also a distinct dorsal groove. The limbs are rather weak and widely separated. The inner digits are rudimentary.

The color above is a light brownish red; below, a pale salmon. The species closely resembles some of the varieties of *Spelerpes ruber*, and in cases of doubt it may be necessary to expose the premaxillary bone in order to determine with certainty the genus.

Spelerpes ruber, (Daudin.).
Red Triton.

Salamandra rubra, Daudin, 1802, *69*, viii, 227; *Spelerpes ruber*, Boulenger, 1882, *28*, 62; Cope, 1889, *51*, 172, with figures.

This species has not at this date been found within the territory of Indiana; yet since it has been taken about Cincinnati and Columbus, Ohio, and in the region about Chicago, it undoubtedly is an inhabitant of Indiana. It is, therefore, proper that it should be here described, for the double purpose of calling attention to it and of enabling observers to identify it. Wherever it occurs it appears to be an abundant species.

This is a rather heavily built species, although the young are more slender. The width of the head is contained in the distance to the groin six times or a little less. The length of the tail two-fifths the total length. It is thick at the base, flattened toward the tip, keeled above from a short distance from the base, and below along the posterior half. The limbs are feebly developed; when the fore and hind legs are pressed to the side they lack about seven costal interspaces of meeting. Of the costal grooves there are fifteen, rarely sixteen. The tongue is boletoid. The vomerine teeth run along the hinder border of the choanæ transversely, then bend at a right angle and run backward to meet the parasphenoidal patches.

The color above varies from salmon to orange, vermillion, or brownish red. The belly is usually of some shade of red and without spots. Above there are numerous spots of black, which are distinct in the young, but in the old are more diffused, giving a brownish tinge to the whole upper surface. The total length may be six inches.

This beautiful animal is more aquatic than any others of the genus that we have with us. Prof. Cope states that its chief haunts are cold springs, but it is frequently seen in damp situations under the bark of fallen trees. It comes on the land after rains.

Spelerpes guttolineatus, (Holbrook.)
Holbrook's Triton.

Salamandra guttolineata, Holbrook, 1842, *54*, v, 29, pl. 7; *Spelerpes guttolineatus*, Boulenger, 1882, *28*, 65; Cope, 1889, *51*, 170, with figures.

This is another species that may yet be found to occur in Indiana. It is most abundant in the Alleghany Mountains, but has been reported by Robert Kennicott from New Madrid, Mo.

It is of the size and general proportions and features of *Spelerpes longicaudus*. It has, however, thirteen instead of twelve costal grooves. Above, the color is brownish yellow. There are three stripes of dark brown, one in the median line, another on each side, beginning at the eyes and extending to the tip of the tail. The under surface is yellow, mottled with brown. Professor Cope says that its habits resemble those of *Plethodon cinereus*. It appears, then, to be essentially terrestrial.

Hyla carolinensis, (Pennant).
Carolina Tree-frog.

Calamita carolinensis, Pennant, 1792, *59*, ii, 331; *C. cinerea*, Schneider, 1799, *41*, i, 174; *Hyla viridis*, Holbrook, 1842, *54*, iv, 119, pl. 29; *Hyla carolinensis*, Günther, 1868, *60*, 105; Boulenger, 1882, *27*, 377; Cope, 1889, *51*, 366, with figures.

This species has not been seen in Indiana up to this time. It appears to be a common frog in the Southern States from Washington, D. C. (W. P. Hay) to Texas, but it has been taken at St. Louis, Mo., by Dr. Englemann; and Prof. Harry Garman (*61*, 189) reports having found it in Union county, Illinois. It is, therefore, to be looked for in the southern portion of our State.

The size is usually greater than that of *H. versicolor*, the head and body being sometimes more than two inches. The head tapers to the rather pointed snout, and is rather longer than broad. The body is slender and the limbs long. The heel reaches in front of the eye. The fingers are distinctly webbed; the toes are webbed to the disks. The digital disks are a little smaller than the tympanum. The surface above is smooth or faintly granulated. The belly and lower parts of thighs are strongly granulated; the throat moderately so. The color above is brownish or olive-green in spirits, but grass-green in life. There are some spots of a light color; at the snout starts a streak of white which runs along the upper lip, under the ear, and over the arm to the side of the body. A similar streak runs along the hinder border of the arm, and still another along the hinder border of the tibia and foot.

Rana septentrionalis, Baird.
Northern Frog.

Rana septentrionalis, Baird, 1855, *1*, 51; Garnier, 1883, *22*, 945; Boulenger, 1882, *27*, 37; Cope, 1889, *51*, 416, with figures.

Rana septentrionalis is a species of frog which ranges from New York to Minnesota, and which may therefore be expected to occur in Northern Indiana. It resembles closely *R. clamata*. The patches of vomerine teeth, however, are minute and do not project behind a line joining the posterior borders of the choanae. Tympanum variable, not as large as the eye. Skin above often rough, but without pustulations. Two dorso-lateral glandular folds, as in *R. clamata*, but sometimes indistinct. Feet fully webbed ; soles without tubercles. Above, the color is light or dark olive, with many narrow, irregular lines of paler. The sides and hinder half of the back with large spots of brown. The hind limbs with dark bars. The living frog is said to have a peculiar minky odor (*22*, 1883, 945).

Rana cantabrigensis, Baird.
Cambridge Frog.

Rana cantabrigensis, Baird, 1854, *1*, 62 ; Boulenger, 1883, *27*, 45 ; Cope, 1889, *51*, 435, with figures.

Rana cantabrigensis is a near relative of *R. sylvatica*, and ranges through British America and Alaska. Not many specimens of it appear to have been taken within the United States, but in the National Museum is a specimen (*51*, 437) which is reported to have come from Clark County, Illinois. This county bounds Vigo County, Indiana, on the west, and it is probable that this frog also belongs to Indiana territory. It is more likely to be found in the northern portion of the State. Specimens of supposed *sylvatica* need to be closely scrutinized in search of *cantabrigensis*.

This species differs from *sylvatica* in having the heel reach only to the middle of the orbit, the tympanic disk only half the diameter of the orbit, the skin of the back between the glandular folds smooth, three phalanges of longest toe free from web, and with a light stripe down the middle of the back. Probably no one of these characters will be sufficient to identify this frog, but most of them must be present.

Carphophis vermis, (Kenn.).
Worm Snake.

Celuta vermis, Kennicott, 1859, *1*, 99 ; *Carphophiops vermis*, Cope, 1875, *12*, 34 ; *Carphophis vermis*, Smith, 1882, *18*, 699 ; *Carphophis amœna* var. *vermis*, Garman, 1883, *13*, 101.

Carphophis vermis is found from St. Louis, Mo., to Kansas and south to Louisiana. It is to be expected in Southern Indiana. It is regarded by Garman as only a variety of *C. amœna*. There are two pairs of frontals, as in *amœna*. However, the head is smaller and the snout more pointed than in *amœna*. The color above is dark iridescent purple, almost black. This dark color descends only to the third row of dorsal scales and ceases abruptly. The upper lip is of the color of the abdomen. This is flesh color ; while the tail below is red. In a specimen taken at Little Rock, Ark., there are 145 ventral plates and 25 subcaudals. The length is 10 inches. I regard this form as a good species.

Virginia valeriæ, B. & G.

Valeria's Snake.

Virginia valeriæ, Baird and Girard, 1853, *6*, 127; Garman, 1883, *13*, 98, pl. vii, fig. 3.

This species is said to occupy the territory from Maryland to Georgia and Illinois. It is therefore likely to be found in Indiana. Specimens in the National Museum are reported to be from Cook county, Ill., and from "Southern Illinois."

It differs from *V. elegans* especially in having the scales in only 15 rows, instead of 17. Ventral plates 118 to 127; subcaudals 24 to 36. Length 11 inches or less. Color yellowish or grayish brown; dull yellow beneath. On the upper surface there are a few black dots; while on each scale of the back there is a faint line, which makes the body appear as if striated. Named for Miss Valeria Blaney, of Maryland.

Abastor erythrogrammus, (Daudin).

Red-lined Snake.

Abastor erythrogrammus, Baird and Girard, 1853, *6*, 125; *Hydrops erythrogrammus*, Garman, 1883, *13*, 35.

The genus *Abastor* differs from the genus *Farancia* only in having the prefrontals distinct, those of opposite sides not being fused. *Abastor erythrogrammus* reaches a length of 40 inches, and has the same colors as *Farancia abacura*. The red, however, forms three longitudinal lines, or stripes. The two lower, or outer, rows of scales are red. Occasionally this stripe is a little wider or a little narrower. The next three rows of scales are black or blue-black. Then follows a stripe of red, occupying one row of scales. Above this is another blue-black stripe three scales in width. The tenth row of scales, the one on the middle of the back, is red. The lower surface is red or flesh color, with a series of blackish spots near the ends of the ventral plates. There are some spots of red on the head. Thus we have here a striped snake, instead of a spotted one, as is *Farancia abacura*. There are about 180 ventral plates.

This snake is best known as a resident of the Southern States east of the Alleghany Mountains, but it has been reported, on the authority of Mr. Garman (*13*, 35) and others as having been taken about Mt. Carmel, in Illinois. If this is true, it will yet be found in Indiana.

Genus HALDEA, B. & G.

Haldea, Baird and Girard, 1853, *6*, 122.

A genus containing a single species of small snakes. Head rather elongated, hardly distinct from the neck; the snout pointed. The place of the prefrontals occupied by a single small plate. Postfrontals large, entering the orbits and suppressing the anteorbitals. Loral present. Nasals 2. Scales keeled. Anal plate divided.

Haldea striatula, (Linn.).

Brown Snake.

Coluber striatulus, Linnæus, 1766, *64*, 375; *Haldea striatula*, Baird and Girard, 1853, *6*, 122; Cope, 1892, *3*, xiv, 376; *Virginia striatula*, Garman, 1883, *13*, 97, pl. vii, fig. 2.

A small, slender snake, not exceeding probably one foot in length; the tail forming about a fifth the total. Head pointed. Crown shields 8, the prefrontals being united. Nasals 2, with the nostril between. A single postorbital and one large temporal. Inferior labials 6. Superior labials 5, the eye over the 3d and 4th. Scales in 17 rows, feebly keeled, those of one or two outer rows large and smooth. Ventral plates 119 to 133; pairs of subcaudals 25 to 46.

Color of upper surface grayish or reddish brown. Abdomen yellowish. There is said sometimes to be a light chestnut band across the back of the head.

Distribution chiefly in the Southern States; said by Prof. S. Garman to occur from Massachusetts to Mississippi. I have taken it in central or southwestern Arkansas.

Little or nothing is known concerning the habits of this snake. In a female taken in Arkansas I found about half a dozen long slender eggs, in each of which was found a young *Haldea*. From the thinness of the egg membranes I judge that the young are brought forth alive and active.

Natrix rigida, (Say).

Stiff Snake.

Coluber rigidus, Say, 1825, *2*, 239; *Tropidonotus rigidus*, Holbrook, 1842, *54*, iv, 39, pl. 10; *Regina rigida*, Baird and Girard, 1853, *6*, 46; *Tropidonotus leberis*, var. *rigidus*, Garman, 1883, *13*, 28; *Natrix rigida*, Cope, 1892, *3*, 668.

This species has been reported from Indiana, but there is now so much doubt about the matter that I prefer to arrange it among the species that have not yet been taken in the State. Indeed it is doubtful if it occurs west of the Alleghany Mountains.

The scales are arranged in 19 rows. Ventral plates 132 to 135; subcaudals 52 to 71. The outer row of scales is smooth, the second nearly so. The ground-color is greenish brown. The middle of the back is of the ground-color, but on each side of this there is a narrow dark stripe, which centers on the 8th row of scales. These stripes are usually distinct. Along the flanks, lying on the 2d row of scales, is a band of pale color. The belly is yellowish, with two rows of dark spots, which lie close together and run from the throat to the vent. Upper jaw, lower jaw and throat yellow.

A specimen with 170 ventrals, from Illinois, on which, in Jordan's Manual, I assigned *rigida* to that State, seems to be rather a specimen of *N. grahamii*.

Natrix grahamii, (B. & G.).

Graham's Water Snake.

Regina grahamii, Baird and Girard, 1853, *6*, 47; *Tropidonotus grahamii*, Günther, 1858, *26*, 78; *Tropidonotus leberis*, var. *grahamii*, Garman, 18˙3, *13*, 28; *Natrix grahamii*, Cope, 1892, *3*, 668.

This species occurs from Michigan to Louisiana and Texas. It is therefore certain to be found in Indiana, though no one has yet recorded it.

The rows of scales are 19, and all keeled; and the number of ventral plates is greater than in the related species, being from 160 to 170. Pairs of subcaudals about 60.

The general color is brown. The middle of the back has a pale band occupying 3 scales in width. Outside of this, on the 8th row of scales, is a narrow dark stripe. On each flank, occupying the three outer rows of scales, is a pale streak. Both above and below this is a very narrow dark stripe. The belly is uniform yellow, except that in the middle line, along the hinder portion of the body, there is a dusky stripe.

The eggs of *N. grahamii* (taken from the oviducts) were found to be 1.5 inch long by .75 in the short diameter. The young, taken from such an egg, was 7 inches long, and was partly enclosed in a sort of concave disk formed of the remains of the yolk. The stripes of the young were much more distinct than those of the adult, and the determination of the species was less difficult. The young are no doubt brought forth alive and active.

Natrix cyclopion, (Dum. & Bib.)

Florida Water-snake.

Tropidonotus cyclopion, Duméril & Bibron, 1854, *74*, vii, 576; Garman, 1883, *13*, 26, pl. 2, fig. 4; *Natrix cyclopium*, Cope, 1892, *3*, 673.

Natrix cyclopion is best known as a Floridan snake; but it has been sent to the National Museum, by Robert Kennicott, from Southern Illinois, and Mr. Garman, as cited above, gives its range as "Ohio to Florida." It is, therefore, not unreasonable to expect to find it yet in Indiana.

The scales of *N. cyclopion* are disposed in from 27 to 33 rows. All the scales are distinctly keeled, except those of the outer row, which are only faintly keeled. The ventrals are from 138 to 150; the subcaudals, 60 to 75 pairs. There is one anteorbital, two postorbitals and two suborbitals. The species may usually be readily recognized by the upper labials being cut off from contact with the eye by these suborbitals. However, this test occasionally fails. The general color is an olive brown or a lead color. Along the sides are faint vertical bands two scales long, and separated by narrow spaces. These may meet on the back, or they may alternate. This species appears to be closely related both to *N. taxispilota* and *rhombifera*.

Ophibolus calligaster, (Harl.).

Coluber calligaster, Harlan, 1835, *39*, 122; *Ophibolus evansii*, Kennicott, 1859, *1*, 99; *Ophibolus calligaster*, Cope, 1875, *12*, 37; *Ophibolus triangulus*, var. *calligaster*, Garman, 1883, *13*, 66.

Resembles *O. doliatus triangulus*, but has the ground-color lighter. The scales are in 25 or occasionally 27 rows. Light olive-brown or gray, with a dorsal series of about 60 subquadrangular, emarginate, dark chestnut-brown blotches, and on each side two series of smaller lateral spots. Belly reddish yellow. Central Illinois to Indian Territory. Dr. Yarrow (*10*, 94) reports this snake in the National Museum from Mt. Carmel, Ill. There is, therefore, no reason why it should not occur in Indiana. It has been stated to occur at Brookville, but since there is doubt about the correctness of the identification, I do not include it among the known snakes of the State.

Genus PITUOPHIS, Holbrook.

Pituophis, Holbrook, 1842, *54*, iv, 7; *Pityophis*, Hallowell, 1852, *1*, 181; Baird and Girard, 1853, *6*, 64; Garman, 1883, *13*, 51.

Large and rather stout snakes, with distinct heads and short tails. Head rather long and pointed. Crown shields 9 or 11; the postfrontals being usually divided into a transverse row of four. Rostral high and narrow. Loral present, small. Anteorbitals 2, rarely 1. Postorbitals 3 or 4. Nasals 2, with the nostril between. Scales keeled; in 29 to 35 rows. Anal plate entire.

Pituophis melanoleucus, (Daudin.)
Eastern Bull-snake. Pine Snake.

Coluber melanoleucus, Daudin, 1802, 69, vi. 409; *Pituophis melanoleucus*, Holbrook, 1842. 54, 7, pl. 1; Baird and Girard, 1853, 6, 65.

This snake has an Eastern and Southern range, but it has been reported by Dr. W. H. Smith (18, 688), on what authority I do not know, to occur occasionally in the State of Ohio. It is possible, therefore, that it may yet be found along our eastern border. Hence I give a description of the species.

The head is broad behind, pointed in front. Two pair of postfrontals, an inner and an outer. Rostral narrow, rising to the inner postfrontals. One anteorbital. Upper labials 7, the 4th entering the orbit. Lower labials 13 or 14, the 5th and 6th largest. Scales in 27 to 29 rows, all keeled except about 4 of the lower rows. Ventral plates 212 to 216; subcaudals, about 60 pairs.

The ground-color above is whitish. On the middle of the back there is a series of about 30 large reddish brown blotches, all margined with black. On the sides there are three or four series of irregular spots, more or less indistinct. Posteriorly these unite into transverse bands, which may alternate with the dorsal series or come opposite them. The abdomen is yellow, with a series of brown blotches on each end of the ventral plates. The length may become five feet or more.

Pituophis sayi, (Schleg.).
Western Bull-snake. Say's Pine Snake.

Coluber sayi, Schlegel, 1837, Essai, ii, 157; *Pituophis sayi*, Baird and Girard, 1853, 6, 152.

This species is much more likely to be found in Indiana than the one just described. In some parts of Illinois it was, in the past, quite abundant, and it has been found well into the eastern part of that State. I have no record of its occurring in Indiana, but it may be looked for along the western border, especially northward, with confidence.

It differs from *melanoleucus*, especially in its coloration. The spots along the back are more numerous, being about 70 in number. The general color, too, is darker, being reddish yellow or, on the back, chesnut brown. The ventrals are also greater in number than in *melanoleucus*, ranging from 220 to 240.

Genus TROPIDOCLONIUM, Cope.

Tropidoclonium, Cope, 1860, 1, 76.

Head little distinct from the body. Crown shields 9. Loral present. Nasal single, grooved below the nostril. Anteorbital 1. Postorbitals 2. Scales in 19 rows; keeled. Anal plate entire.

Tropidoclonium lineatum, (Hallowell).

Streaked Snake.

Microps lineatus, Hallowell, 1856, *1*, 241; *Tropidoclonium lineatum*, Cope, 1860, *1*, 76; H. Garman, *61*, iii, 187; *Storeria lineata*, S. Garman, 1883. *13*, 32, pl. 1, fig. 4.

This species will almost certainly yet be taken within the limits of our State. It has been found lately by Prof. Harry Garman, at Urbana, Ills., within 35 miles of the Indiana line. It has been reported from Hughes, Ohio, by Dr. Yarrow (*10*, 131), but I am informed by Dr. L. Stejneger, of the National Museum, that the specimen so determined is a *Storeria*.

This species greatly resembles *Tropidonotus rigidus*, but the generic characters are quite different. The scales are in 19 rows, are strongly carinated, and the animal has a very rough appearance. The ventrals are from 138 to 150; the subcaudals 26 to 35 pairs. The 6th upper labial has been crowded from its place and lies above and between the 5th and 7th. The ground-color above is ashy-brown. Along the middle of the back there is a streak of gray one scale and two half scales wide, and a similar band along each side on the 2d and 3d rows of scales. A row of black spots on the bases of the lowest row of scales, another just above the lateral line, and a third on each side of the dorsal streak. Belly gray or yellow, with two rows of black spots, one on each side of the middle line, the two spots on many of the ventrals narrowly confluent.

The size is somewhat greater than that attained by *Storeria dekayi*. The distribution of the species is from Texas to Kansas and east to Illinois.

Agkistrodon piscivorus, (LaC.).

Water Moccasin; Mokason.

Crotalus piscivorus, LaCépède, 1789, Serp. ii, 130; *Trigonocephalus piscivorus*, Holbrook, 1842, *54*, iii, 33, pl. 7; *Toxicophis piscivorus*, Baird and Girard, 1853, *6*, 19; *Ancistrodon piscivorus*, Garman, 1883, *13*, 121, pl. 8, fig. 2.

This is an abundant snake everywhere about the water courses of the Southern States. It has not yet been seen by any scientific observer in any locality in Indiana, although Mr. Robert Ridgway says that he has been informed that it was abundant at one time about Vincennes. I believe that it will yet be found along the lower portions of the Wabash river and the neighboring parts of the Ohio. It is found in numbers in the swamps in Union county, Illinois, and this is little further south than

Posey county, Ind. However, no specimens of it were in the collection of Mr. Sampson, of New Harmony. It ought to be diligently sought for in that region, and specimens saved. It is entirely aquatic and will be found hiding among stones, and in fallen timber, or basking in the sun along the banks of ponds and streams.

This reptile differs from the Copperhead in having no loral plate, no suborbitals, the scales in 25, instead of 23 rows, and colors darker. The ground color is dark chestnut-brown, with blotches of still darker brown. The head is very dark. Upper lip with a whitish streak that continues back on the neck. Belly yellowish, with many blotches of black.

Aromochelys carinata, Gray.

Keeled Mud Turtle.

Aromochelys carinata, Gray, 1855, *25*, 47, pl. 20; Yarrow, 1875, *19*, v, 582; *Goniochelys triquetra*, Agassiz, 1857, *4*, i, 423; *Cinosternum carinatum*, Boulenger, 1889, *84*, 38.

This species is common in the streams and ponds of the Southern States from Georgia to Arizona. It has been reported from Northern Illinois by Messrs. Rice and Davis, and if their determination is correct it adds to the known range of this turtle in a remarkable manner, and as a consequence it is to be looked for in Indiana along the Wabash River. It differs from *A. odorata* in attaining a larger size, in having a larger head and stronger jaws, and in having a high shell with a median sharp keel toward which the slightly convex sides slope up roof-like. The scutes of the carapace overlap those lying behind them. The plastron is truncated in front and notched behind. The gular scute is wanting. The pectorals are large, and the suture between them is longer than that between the humerals. The males are furnished with two patches of small, sharp-edged scales, one above, the other below, the hollow of the knee. The color of the shell and skin above is olive, with streaks of yellow and spots of brown. The posterior borders of the scutes of the carapace are blackish. All the inferior surfaces are more yellow. There are no streaks of yellow on the head. Habits in general those of its relative, the Musk Turtle.

Chrysemys picta, (Schneider).

Painted Tortoise.

Testudo picta, Schneider, 1783, *87*, 348; *Emys picta*, Holbrook, 1842, 54, i, 75, pl. 10; *Chrysemys picta*, Gray, 1855, *25*, *32*; Agassiz, 1857, *4*, i, 438, pl. i, figs. 1-5, and pl. iii, fig. 4; Boulenger, 1889, *84*, 72.

This species has at various times been reported from localities in this State, and was given as a resident of Indiana in my "Preliminary List."

There is, however, now so much doubt concerning it that I give it as merely one of the possibilities and leave it to the future to settle.

C. picta is a close relative of *marginata*, so close that some authors regard the latter as only a variety of the former. *Picta* differs from *marginata* principally in the arrangement of the vertebral and costal scutes, these being so disposed that the sutures between the costals meet the corresponding sutures between the vertebrals. Hence the scutes form four straight rows right across the back, an unusual thing among turtles. The red or yellow margin of the front of the scutes is also much wider than in *marginata*, and they produce conspicuous colored bands across the shell. In the collection of the National Museum I have seen a specimen with the plates somewhat intermediate in arrangement; nevertheless I believe that *picta* has quite thoroughly and in a remarkable manner freed itself, as a species, from *marginata*.

HABITS.—The habits of *C. picta* are similar to those of *C. marginata* and have been more carefully observed. The reader is referred to Agassiz's "Contributions" for further information. Prof. J. A. Allen says of these turtles, as known to him in Massachusetts: "The shrill, piping notes of this species is frequently heard in May and June, especially during intervals between showers on hot sultry days."

Chrysemys belli, (Gray).

Emys belli, Gray, 1831, *112*, 31; *Chrysemys belli*, Gray, 1855, *25*, 33; Agassiz, 1857, *4*, i, 439, pl. 6, figs. 8-9; *Chrysemys cinerea*, var. *belli*, Boulenger, 1889, *84*, 74,

This is another species that has not yet been taken within our limits, but which may nevertheless be an inhabitant of the State. It occurs from Mississippi to western Illinois, thence northwestward to British Columbia. It has been taken in western Illinois by Agassiz, and more recently by Prof. Harry Garman, and it appears to be very common in the sloughs of the "bottom land" about Quincy. It should be looked for in the lower Wabash.

The arrangement of the scutes of the carapace is the same as in *C. marginata*. The ground-color is copper-red or bronze-color. Across the costal scutes there are some irregular red or yellow bands and some red dots. The marginals are divided above by a yellow streak, while on their lower surface there are black, eyelike spots on a red ground. The plastron is covered all over with blackish markings of various shapes. The size becomes greater than that of either *picta* or *marginata*, some 8 inches as the length of the shell.

Cistuda ornata, (Ag.).

Ornate Box-tortoise.

Cistudo ornata, Agassiz, 1857, *4*, i, 445, pl. iii, figs. 12 and 13; Boulenger, 1889, *84*, 118.

This species of the genus *Cistuda* occurs abundantly west of the Missouri River, being, as reported by F. W. Cragin, "so abundant in some sections of Southern Kansas that it amounts to a nuisance as a cumberer of the ground." (*49*, 100.) The species, however, ranges further east into Illinois, and some years ago specimens were sent to the National Museum from Fairfield, Wayne County, Illinois, within 35 miles of the Wabash River (*10*, 37), hence its occurence within Indiana territory may be discovered at any time.

The shell of *ornata* is proportionally shorter and broader than that of *carolina*. The back is also flatter, especially along the middle line, and there is no trace of a keel, even in the young. The plastron does not completely close the shell, and there is a short but distinct bridge. The head is larger, the snout shorter, and the upper jaw is notched in front. The shell is more elegantly marked than in *carolina*, the ground color being olive-brown to black, while there are numerous spots and streaks of bright yellow. The yellow markings usually seem to radiate from the center from which each scute began to grow.

GLOSSARY.

ABDOMINAL SCUTES, or ABDOMINALS (of tortoises). See figure 12.
ALVEOLAR SURFACE. A flat masticatory surface of the jaws of tortoises, seen just within the cutting edge.
AMPHICŒLOUS. Concave at both ends; said of certain vertebræ
ANAL PLATE. The large scale immediately in front of the vent of serpents.
ANKYLOSED. Joined by bony union.
ANTEORBITAL. A small epidermal plate of the head of snakes, which lies immediately in front of the eye. If there are but three plates between the eye and the nostril, either the anteorbital or the loral is missing. If the plate present next the eye has its greatest length horizontal, it is the loral; otherwise it is the anteorbital.
AXILLA. The arm-pit.
AXILLARY. Pertaining to or placed in the axilla.
AZYGOUS. Placed in the middle line, and, therefore, without a fellow.

BALANCERS. Organs of adhesion at the sides of the mouths of tadpoles; "holders."
BARBELS. A short, worm-like process of skin about the mouth or at the chin.
BRANCHIÆ. Gills.
BRANCHIAL ARCHES. Bony or cartilaginous arches that support the gills of fishes, or arches that correspond to these in other animals.
BRIDGE. That portion of the shell of a tortoise which joins the carapace and the plastron.

CALLOSITIES. Patches of hard skin on the plastron of soft-shelled turtles.
CANTHUS ROSTRALIS. A slight ridge from the eye to the tip of the snout, separating the upper surface of the head from the side.
CARAPACE. The upper portion of the shell of tortoises.
CARINATED. Furnished with a keel, or sharp ridge.
CHOANÆ. The internal nostrils.
CLAVICLES. A bone corresponding to the human collar-bone.
CLOACA. The common chamber into which the intestine, the ureters and the genital ducts open.

CORACOID. The bone or cartilage that forms the portion of the shoulder girdle in front of the glenoid cavity.

COSTAL. Pertaining to the ribs. The costal furrows, or grooves, of Urodeles run across the body between the fore and the hind legs. For costal scutes of tortoises, see figure 11.

CROWN-SHIELDS. The large plates which cover the upper surface of snakes' heads. See figure 9.

DENTARY. The anterior bone of the lower jaw, the one usually bearing the teeth.

DERMAL FOLDS. The thickened ridges of skin on the backs of some frogs; the glandular folds.

DIAPOPHYSIS. The transverse process of a vertebra; here used of that of the sacral vertebra.

ENTOPLASTRON. One of the bones of the plastron of a tortoise.

EMARGINATE. Furnished with an obtuse notch.

EMARGINATION. A broad, shallow notch.

EPIDERMAL. Belonging to the outer layer of the skin.

ETHMOID. A bone or cartilage between and often surrounding the nasal sacs.

EUSTACHIAN TUBE. A passage from the tympanic cavity to the pharynx.

EXTRALIMITAL. Outside of the limits immediately adjoining Indiana.

FEMORAL. Pertaining to the thighs; femoral pores are found on the under side of the thighs of some lizards.

FRONTALS. Plates of the heads of snakes. See figure 9.

FONTANELLE. A space filled with membrane between bones that approach one another without meeting.

GULAR. Pertaining to the throat; gular fold, a fold of skin across the throat; gular scute (of tortoises), see figure 12.

HOLDERS. Organs of adhesion at the corners of the mouth of a tadpole.

HUMERALS. Scutes of the plastron of a tortoise; see figure 12.

INFRAMARGINALS. Scutes of some tortoises lying below the marginals.

INGUINAL. Pertaining to the groin.

KEEL. A well defined ridge. Keeled, furnished with a sharp ridge.

LABIAL. Pertaining to the lips. Labial plates (of snakes), see figure 9.

LARVA. The undeveloped young of some animals, as the caterpillar of an insect or the tadpole of a frog.

LORAL. Pertaining to the space in front of the eye; see anteorbital, also figure 9.

MARGINALS. See figure 11.

MAXILLARY. A bone of the upper jaw lying behind the premaxillary of each side, usually bearing the outermost row of teeth; missing in *Siren* and *Necturus*.

METAMORPHOSIS. Transformation; change from tadpole to adult form.

NUCHAL PLATE. See figure 11.

OCCIPITAL. Belonging to the hinder part of the head. See figure 9.

OCELLATED. Furnished with eye-like spots, spots consisting of concentric rings.

OLIVE. Brownish or yellowish green.

OPISTHOCŒLOUS. Said of vertebræ which are concave at the hinder end and convex at the anterior end.

OVIPAROUS. Producing young from eggs that hatch after deposition.

OVOPOSITION. The laying of eggs.

OVOVIVIPAROUS. Producing young from eggs which hatch before being laid.

PALATINE. A bone of the roof of the mouth lying behind the vomer on each side.

PALATOPTERYGOID. The united palatine and pterygoid.

PAPILLA. Minute folds or elevations of the skin or mucous membrane.

PARASPHENOID. A broad bone underlying the brain-case; parasphenoidal teeth are found in the hinder part of the roof of the mouth.

PARATOID. Belonging to the region found at the hinder lateral part of the head; paratoid glands of toads are elevated glandular bodies at the sides of the back part of the head.

PECTORALS (of tortoises). See figure 12.

PINNATE. Arranged along the sides of a central axis like the vane of a feather.

PLASTRON. The lower portion of the shell of a tortoise.

PLEURODONT. With the teeth grown fast to the inner side of the bone of the jaw.

POST-FRONTALS. See figure 9.

PREFRONTALS. See figure 9.

PREMAXILLARY. The bone forming that part of the upper jaw immediately at the snout. The two premaxillaries are sometimes united.

PROCŒLOUS. Said of vertebræ which have the anterior end concave, the hinder convex.

PTERYGOID. A bone of the roof of the mouth lying on each side immediately behind the palatine.

QUADRATE. The bone on each side to which the lower jaw of batrachians and reptiles is swung.

ROSTRAL. The epidermal plate covering the snout of snakes and lizards. See figure 9.

SCUTE. A large epidermal scale.
SEPTUM. A dividing wall, as that between the nasal passages.
SNOUT. The portion of the head in front of the eyes.
SPLENIAL. A bone at the hinder and inner part of the lower jaw; occasionally bearing teeth.
SQUAMOSAL. A bone usually overlying the inner ear; in snakes attached to the hinder portion of the skull and supporting the quadrate.
SUBCAUDALS. The large scales on the underside of the tail of a snake.
SUPERCILIARIES. The plates over the eye of a snake. See figure 9.
SUPRAORBITALS. Same as superciliaries.
SYMPHYSIS. The union of two bones of opposite sides, in the middle line, and with little or no motion.
TAIL. Portion of the body behind the vent.
TEMPORAL ARCH. A bony bar from the upper jaw to the quadrate, overlying the temporal muscle; found in some tortoises.
TERETE. Cylindrical and tapering.
UROSTYLE. The rod-like posterior termination of the spinal column of frogs.
VENT. The opening outwardly of the cloaca.
VENTRAL PLATES. The epidermal plates on the belly of snakes, etc.
VERTEBRAL SCUTES. The median row of plates on the back of a tortoise. See figure 11.
VERTICAL. See figure 9.
VILLIFORM. Having the form or appearance of villi; like the pile of velvet.
VITTA. A stripe.
VOMER. A bone lying in the roof of the mouth just behind the premaxillary. One on each side of batrachia and reptiles.
VOMERO-PALATINE. The united vomer and palatine.

List of publications referred to in this work, together with the numbers by which they are cited.

1. Proceedings of the Academy of Natural Sciences of Philadelphia.
2. Journal of the Academy of Natural Sciences of Philadelphia.
3. Proceedings of the U. S. National Museum, Washington, D. C.
4. Agassiz's Contributions to the Natural History of the United States. 1857.
5. Wilkes' United States Exploring Expedition: Herpetology. 1858.
6. Baird and Girard's Serpents of North America. 1853.
7. Popular Science Monthly.

8. Bulletins of the U. S. Geological and Geographical Survey.
10. Yarrow's Check-list of N. A. Reptilia and Batrachia. 1882.
12. Cope's Check-list of N. A. Batrachia and Reptilia. 1875.
13. Samuel Garman's Serpents of North America. 1883.
14. Long's Expedition to the Rocky Mountains. 1823.
18. Reptiles and Amphibians of Ohio. Dr. W. H. Smith in Geological Survey of Ohio for 1882.
19. Wheeler's Survey West of 100th Meridian. Vol. V, 1878.
20. Reports on the Pacific Railroad Survey. Vol. X, 1859; Vol. XII, 1860.
22. American Naturalist.
25. Catalogue of the Shield Reptiles of the British Museum, 1855; Supplement, 1870; and Appendix, 1872. Dr. John Edward Gray.
26. Catalogue of the Colubrine Snakes of the British Museum. Dr. A. Günther. 1858.
27. Catalogue of Batrachia Salientia of the British Museum. Dr. G. A. Boulenger. 1882.
28. Catalogue of Batrachia Gradientia of the British Museum. G. A. Boulenger. 1882.
29. Catalogue of the Lizards of the British Museum. G. A. Boulenger. Vols. I and II, 1885; Vol. III, 1887.
30. Fauna of New York, J. E. DeKay. 1842.
32. Smithsonian Contributions to Knowledge.
34. Proceedings of the American Association for the Advancement of Science.
36. Transactions of the American Philosophical Society.
37. Marcy's Exploration of the Red River Valley, 1854.
39. Medical and Physical Researches. Dr. Richard Harlan. 1835.
40. B. S. Barton's Letter on *Siren lacertina*. 1821.
41. Schneider's Historia Amphibiorum. 1799.
42. Bulletins of the Essex Institute.
43. Proceedings of the Boston Society of Natural History.
47. American Journal of Science and Arts.
48. Bulletin of the American Museum of Natural History. New York.
49. Bulletin of the Washburn College Laboratory of Natural History.
50. Science, a Weekly Newspaper of all the Arts and Sciences. New York.
51. Batrachia of North America. Dr. E. D. Cope. 1889.
52. Fisheries Industries of the United States. Washington, D. C. 1884.
53. Holbrook's Reptiles of North America. Edition of 1838.
54. Holbrook's Reptiles of North America. Edition of 1842.

57. Latreille's Histoire Naturelle des Reptiles. 1802.
58. Storer's Report on the Reptiles of Massachusetts. 1809.
59. Pennant's Arctic Zoology. 1792.
60. Catalogue of the Batrachia of the British Museum. Dr. A. Günther. 1858.
61. Bulletins of the Illinois Laboratory of Natural History.
62. Annals of the Lyceum of Natural History of New York.
63. Reise in das Innere Nord America, von Maximillian Prinz zu Wied. 1839 and 1841.
64. Linnæus' Systema Naturæ. Edition X, 1758; edition XII, 1766.
66. Der Naturforscher.
67. Manual of Vertebrates of the Eastern United States. Dr. D. S. Jordan. Ed. v, 1888.
68. Daudin's Histoire Naturelle des Rainettes, etc. 1803.
69. Daudin's Histoire Naturelle des Reptiles. 1803.
71. Shaw's General Zoölogy. London. 1802.
72. D'Orbigny's Dictionaire d' Histoire Naturelle. 1843.
73. Zoölogical Miscellany. 1842.
74. Duméril & Bibron's Erpétologie Générale. 1834 to 1854.
75. Natürliches System der Amphibien. J. Wagler. 1830.
77. Isis von Oken. Erster Band, 1822.
79. Preliminary Catalogue of the Amphibia and Reptilia of Indiana. O. P. Hay, 1887, in No. 94.
80. American Monthly Magazine.
82. Fitzinger's Systema Reptilium. 1843.
83. Archiv für Naturgeschichte.
84. Catalogue of Chelonians in the British Museum. Dr. G. A. Boulenger. 1889.
85. Annals du Muséum de Paris.
86. Mémoires du Muséum d' Histoire Naturelle de Paris.
87. Schneider's Schildkröte. 1783.
88. Schweigger's Prodromus Monographiæ Chelonorum. 1814.
90. Spix' Species Novæ Testudinum, etc. 1824.
91. J. E. Gray's Catalogue of Tortoises. 1844.
92. Tableau Encyclopedique et Méthodique Erpétologie. Paris. 1789.
93. Reports of the Geological Survey of Indiana.
94. Journal of the Cincinnati Society of Natural History.
95. Amœnitates Academicæ Linnæi.
96. Wagler's Tentamen Systematis Amphibiorum. 1820.
97. DeBlainville's Journal de Physique.
98. Smith's Correspondence of Linnæus. 1821.
99. Tschudi's Classification der Batrachier. 1838.
100. Agassiz' Nomina Systematica Generum Reptilium. 1848.
101. Annals and Magazine of Natural History. London.

102. Contributions to Maclurean Lyceum.
103. Maximillian von Wied's Verzeichniss der Reptilien. 1865.
104. Tijdschrift voor Natuurlijke Geschiedenis en Physiologie.
105. Gravenhorst's Uebersicht der Zoölogischen Systeme. Göttingen, 1807.
106. Siebold's Fauna Japonica. 1838.
107. Atlantic Journal.
108. Rafinesque's Annals of Nature. 1820.
109. Laurenti's Synopsis Reptilium. 1768.
110. Zoölogical Journal.
112. J. E. Gray's Synopsis Reptilium. 1831.
113. Duméril's Catalogue Méthodique des Reptiles. 1851.
114. Annals of Philosophy.
115. Fleming's Philosophy of Zoölogy. Edinburgh, 1822.
116. Bonaparte's Osservazione sulla sec. Ed. del Regne Animale. 1830.
119. Duméril's Zoölogie Analytique. 1806.
120. Schriften der Gesellschaft Naturforschender Freunde. Berlin.
122. Studies from the Biological Laboratory of Johns Hopkins University.
123. Gmelin's Systema Naturæ. 1789.
124. Cuvier's Regne Animal.

PLATES AND DESCRIPTION.

PLATE I.

Figures all four times the natural size.

Figure 1. Open mouth of *Ambystoma microstomum*.
Figure 2. Open mouth of *Plethodon glutinosus*. *V. t.*, the vomerine teeth; *p. t.*, the parasphenoidal bands of teeth. Above the vomerine bands are seen the choanæ.
Figure 3. Open mouth of *Spelerpes longicaudus*, showing the vomerine and parasphenoidel teeth and the tongue.
Figure 4. Roof of the mouth of *Spelerpes maculicaudus*; *v.*, the hook-shaped bands of vomerine teeth; *p.*, the parasphenoidal bands.
Figure 5. Open mouth of *Desmognathus fusca*; *ch.*, the choanæ; *v.*, the vomerine teeth; *p.*, the parasphenoidal teeth; *t.*, the tongue.
Figure 6. Open mouth of *Diemyctylus viridescens*, showing the two rows of vomerine teeth arranged in $\wedge$-form.

PLATE I.

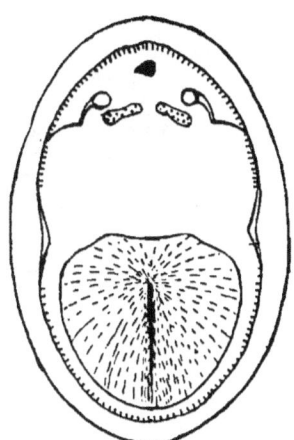

Fig. 1.

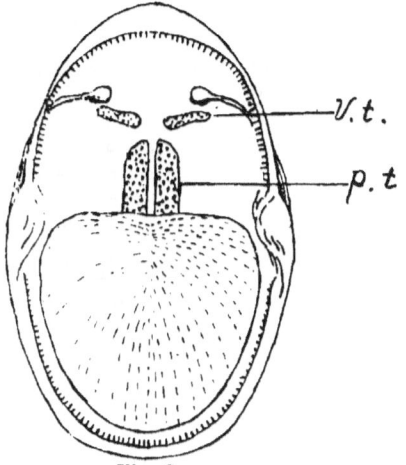

Fig. 2.

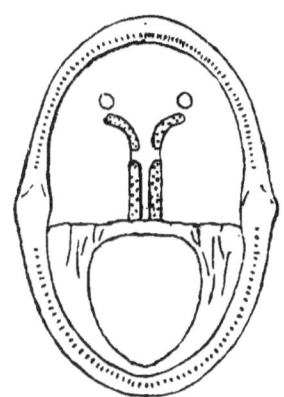

Fig. 3.

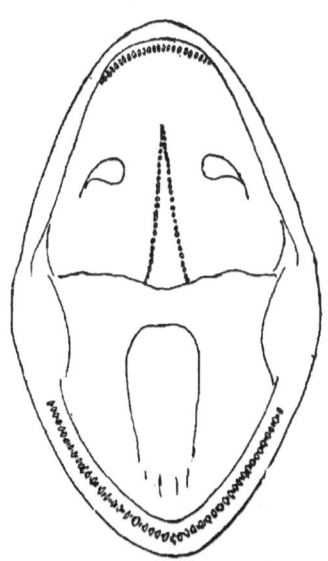

Fig. 4.

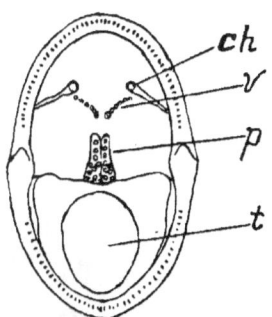

Fig. 5.

Fig. 6.

PLATE II.

Figures all twice the natural size.

Figure 7. Sternum and shoulder-girdle of the Frog, *Rana*. *Os.*, omosternum; *st.*, sternum; *cr.*, coracoid; *cl.*, clavicle; *ec.*, epicoracoid; *sc.*, scapula; *h.*, humerus.

Figure 8. Sternum and shoulder-girdle of Toad, *Bufo*. Letters with the same significance as in Fig. 7.

Figure 9. Upper view of head of Black-racer, *Bascanion constrictor*.

Figure 10. Side view of same head. Numerals with same significance in both figures. 1, vertical plate; 2, occipital plates; 3, superciliaries; 4, the postfrontals; 5, the prefrontals; 6, the anteorbitals; 7, the loral; 8, the postorbitals; 9, the rostral; 10, the upper and lower labials; 11, the temporal; 12, the nasals.

PLATE II.

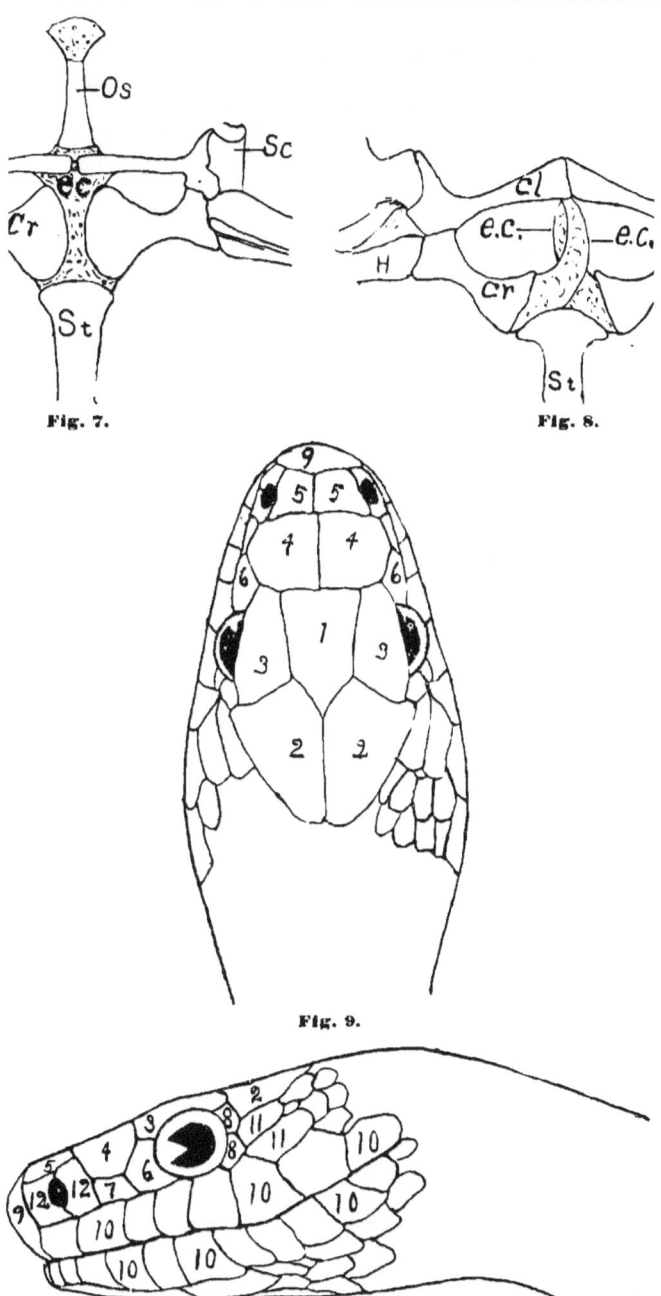

Fig. 7.

Fig. 8.

Fig. 9.

Fig. 10.

PLATE III.

Figure 11. Carapace of *Chrysemys marginata*; *v.*, the neural epidermal plates; *c.*, the costals; *m.*, the marginals; *n.*, the nuchal. Natural size.

Figure 12. Plastron of *Chrysemys marginata*; *g.*, the gular plates; *h.*, the humerals; *p.*, the pectorals; *a.*, the abdominals; *f.*, the femorals; *an.*, the anals. Natural size.

PLATE III.

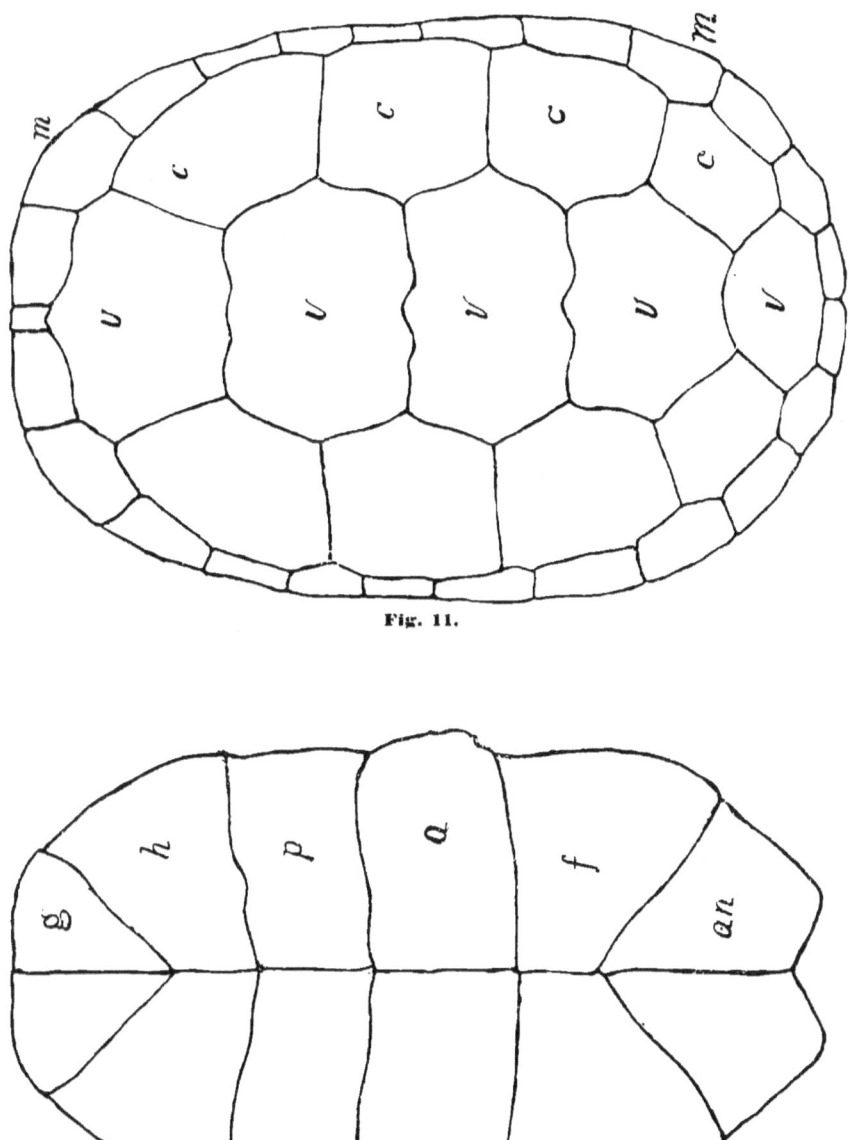

Fig. 11.

Fig. 12.

INDEX.

	Page		Page		Page
Abastor	179	C. nig. triseriatus	62	Garter-snake	116
A. erythrogrammus	179	Chrysemys	156	Geographic terrapin	154
Acris	53	C. belli	187	Glass-snake	131
A. gryllus	53	C. concinna	159	Graham's water-snake	189
A. gryllus crepitans	54	C. elegans	161	Green frog	65
Agkistrodon	123	C. hieroglyphica	158	Green triton	45
A. contortrix	123	C. labyrinthica	158	Ground-snake	75
A. piscivorus	184	C. marginata	163	Gyrinophilus	175
Alleghany black-snake	93	C. picta	185	G. porphyriticus	176
Alleghany salamander	15	C. troostii	160	Halden	186
Alligator lizard	132	Cistudn	171	H. striatula	185
Alligator snapping turtle	151	C. carolina	172	Hellbender	12
Ambystoma	17	C. ornata	187	Hemidactylium	32
A. copeianum	26	Clemmys	168	H. scutatum	32
A. jeffersonianum	22	C. guttata	168	Heterodon	102
A. microstomum	19	Cnemidophorus	137	H. platirhinos	105
A. opacum	29	C. sexlineatus	137	H. simus	107
A. punctatum	27	Coluber	90	Hieroglyphic terrapin	152
A. talpoideum	174	C. guttatus	92	Hog-nosed snake	10
A. tigrinum	23	C. obsoletus	93	Holbrook's triton	176
A. xiphias	174	C vulpinus	90	Hoosier frog	68
Ambystomatidæ	17	Colubridæ	76	Hoosier salamander	39
Amphiuma	12	Congo snake	13	Horn-snake	79
A. means	13	Cope's salamander	26	House-snake	107
Amphiumidæ	12	Copperhead	123	Hyla	55
Anguidæ	131	Coral snake	121	H. carolinensis	177
Appendix	174	Cottonmouth	184	H. pickeringii	58
Aromochelys	153	Cricket frog	53	H. versicolor	56
A. carinata	185	Crotalidæ	122	H. squirella	58
A. odorata	153	Crotalus	128	Hylidæ	52
Ashy salamander	33	C. horridus	128	Iguanidæ	131
Bascanion	81	Cryptobranchidæ	15	Jefferson's salamander	22
B. constrictor	81	Cryptobranchus	15	Joint-snake	134
Black-racer	81	C. alleganiensis	15	Keeled mud-turtle	185
Black-snake	81	Cyclophis	85	King-snake	110
Blanding's tortoise	170	C. vernalis	85	Kinosternidæ	152
Blue-racer	81	Dekay's snake	88	Kinosternon	154
Blue-tailed lizard	140	Desmognathinæ	42	K. pennsylvanicum	154
Box-tortoise	172	Desmognathus	42	Lacertilia	131
Bronzy salamander	175	D. fusca	43	Leather-snake	96
Brown-backed lizard	139	Diadophis	84	Leopard frog	67
Brown snake	180	D. punctatus	84	Lesueur's map-tortoise	165
Brown triton	43	Diamond water-snake	101	Lizards	131
Bufo	49	Diemyctylus	45	Long-tailed triton	38
B. lentiginosus	50	D. viridescens	45	Lygosoma	138
Bufonidæ	49	Eastern mud-turtle	154	L. laterale	139
Bull-snake, Eastern	183	Elapidæ	121	Macroclemys	151
Bull-snake, Western	183	Elaps	121	M. temminckii	151
Cambridge frog	178	E. fulvius	121	Malaclemys	164
Carolina tree-frog	177	Emydoidea	170	M. geographica	166
Carphophis	78	E. bland ngii	170	M. pseudogeographica	165
C. amœna	78	Eumeces	139	Marbled salamander	29
C. vermis	178	E. fasciatus	140	Massasauga	126
Chameleon tree-frog	56	Eutainia	112	Map terrapin	166
Checkered snake	79	E. butlerii	120	Milk-snake	107
Chelonia	142	E. radix	115	Moccasin	184
Chelydra	148	E. saurita	113	Mole salamander	174
C. serpentina	149	E. sirtalis	117	Mud-eel	8
Chelydridæ	148	Farancia	79	Musk turtle	153
Chicken-snake	107	F. abacura	79	Natrix	95
Chorophilus	61	Florida water-snake	181	N. cyclopion	181
C. nigritus	61	Four-toed salamander	32	N. grahamii	181
C. nig. feriarum	62	Fox-snake	90	N. kirtlandi	97

	Page		Page		Page
Natrix leberis	96	R. pipiens	65	S. longicaudus	38
N. rhombifera	101	R. septentrionalis	177	S. maculicaudus	39
N. rigida	180	R. sylvatica	71	S. ruber	175
N. sipedon	98	Ranidæ	64	Spotted coluber	92
Nent terrapin	159	Rattlesnake, Banded	128	Spreading viper	102
Necturus	10	Rattlesnake, Prairie	126	Squirrel tree-frog	58
N. maculatus	10	Red-backed salamander	33	Stiff-snake	180
Newt	45	Red-headed lizard	140	Storeria	88
Northern frog	177	Red-lined snake	179	S. dekayi	88
Ophibolus	107	Red triton	175	S. occipitomaculata	89
O. calligaster	182	Reptiles	74	Storer's snake	89
O. dolintus	107	Ribbon-snake	113	Streaked snake	184
O. dol. triangulus	108	Ring-necked snake	84	Swamp frog	67
O. dol. syspilus	108	Salamandridæ	44	Sword-tailed salamander	174
O. getulus	110	Salientia	48	Teiidæ	136
Ophidia	75	Salmon triton	175	Testudinata	148
Ophisaurus	134	Sand-viper	105	Testudinidæ	155
O. ventralis	134	Scaly salamander	32	Tiger salamander	23
Ornate box-tortoise	187	Sceloporus	132	Toad	50
Painted tortoise, Eastern	185	S. undulatus	132	Tortoises	142
Painted tortoise, Western	163	Scincidæ	138	Trionychidæ	142
Phyllophilophis	86	Siren	8	Trionychoidea	142
P. æstivus	87	S. lacertina	8	Trionyx	143
Pickering's tree-frog	58	Sirenidæ	8	T. agassizii	144
Pine snake	183	Sistrurus	125	T. muticus	143
Pine-tree lizard	132	S. catenatus	126	T. spiniferus	146
Pituophis	182	Spotted salamander	27	Tropidoclonium	183
P. melanoleucus	183	Six-lined lizard	137	T. lineatum	184
P. sayi	183	Slimy salamander	36	Turtles	142
Plethodon	33	Small-mouthed sal'mand'r	19	Two-lined triton	40
P. æneus	175	Smooth green-snake	85	Urodela	6
P. cinereus	33	Snakes	75	Valeria's snake	179
P. glutinosus	36	Snapping turtle	149	Virginia	80
Plethodontidæ	31	Soft-shelled turtle, Agas-		V. elegans	80
Proteidæ	10	siz's	144	V. valeriæ	179
Racine garter-snake	115	Soft-shelled turtle, spine-		Virginia's snake	80
Rana	64	less	143	Water-moccasin	184
R. areolata circulosa	68	Soft-shelled turtle, spiny	146	Water-snake	98
R. cantabrigensis	178	Speckled tortoise	168	Wood-frog	71
Rana catesbiana	70	Spœlerpes	37	Worm-snake	178
R. clamata	69	S. bislineatus	40		
R. palustris	67	S. guttolineatus	176		

www.ingramcontent.com/pod-product-compliance
Lightning Source LLC
Chambersburg PA
CBHW020916230426
43666CB00008B/1467